BIG AND SMALL ANIMALS

FISH

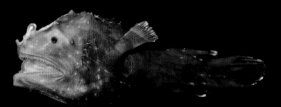

BY BRENNA MALONEY

Children's Press®
An imprint of Scholastic Inc.

A special thank-you to the team at the Cincinnati Zoo & Botanical Garden for their expert consultation.

Library of Congress Cataloging-in-Publication Data
Names: Maloney, Brenna, author.
Title: Fish / by Brenna Maloney.
Description: First edition. | New York: Children's Press, an imprint of Scholastic Inc., 2023. | Series: Big and small animals | Includes index. | Audience: Ages 5–7. | Audience: Grades K–1. | Summary: "Continuation of the Wild World series comparing big and small animal sizes"— Provided by publisher.
Identifiers: LCCN 2022026378 (print) | LCCN 2022026379 (ebook) | ISBN 9781338853537 (library binding) | ISBN 9781338853544 (paperback) | ISBN 9781338853551 (ebk)
Subjects: LCSH: Fishes—Miscellanea—Juvenile literature. | CYAC: Fishes. | Size. | BISAC: JUVENILE NONFICTION / Animals / Fish | JUVENILE NONFICTION / Concepts / Opposites
Classification: LCC QL617.2 .G483 2023 (print) | LCC QL617.2 (ebook) | DDC 597—dc23/eng/20220606
LC record available at https://lccn.loc.gov/2022026378
LC ebook record available at https://lccn.loc.gov/2022026379

10 9 8 7 6 5 4 3 2 1   23 24 25 26 27

Printed in China   62
First edition, 2023

Book design by Kay Petronio

Photos ©: cover top, back cover right, 1 top, 2 right: Pete Oxford/Minden Pictures; cover bottom, back cover left, 1 bottom, 2 left: Danté Fenolio/ Science Source; 4 shark: Andrea Izzotti/Getty Images; 6: Theodore W. Pietsch/University of Washington; 7 right: Danté Fenolio/Science Source; 8–9: Theodore W. Pietsch/University of Washington; 10–11: SeaTops/ Alamy Images; 14–15: Nick Hawkins/Minden Pictures; 16–17: Michael Patrick O'Neill/Alamy Images; 20–21: Phillip Colla/BluePlanetArchive; 22–23: Westend61/Getty Images; 24–25: Doug Perrine/Minden Pictures; 26 shark: torstenvelden/Getty Images; 28–29: Pete Oxford/Minden Pictures; 30 top left: Pete Oxford/Minden Pictures; 30 top right: Danté Fenolio/Science Source; 30 bottom left: Doug Perrine/BluePlanetArchive; 30 bottom right: Rodrigo Friscione/Getty Images.
All other photos © Shutterstock.

PHOTOCORYNUS SPINICEPS

WHALE SHARK

CONTENTS

FISH FACTS

Dive into the world of **fish**! All fish live in water. They breathe underwater using **gills**. They have **scales**, not hair or fur. Most fish have **fins**, especially tail fins that help them swim. Fish are **cold-blooded**. This means their body temperatures change with their surroundings.

Measuring Up

This book is all about size. Which fish are the smallest? Which are the biggest? Why does being big or small matter? An animal's size can determine where and how it lives. Size makes a difference as to what **prey** it can chase and what **predators** it must get away from. You can learn a lot about fish just by their size.

Get ready to discover the different sizes of 10 fantastic fish and why it matters, from the smallest to the biggest!

FACT
Most fish don't have eyelids.

5

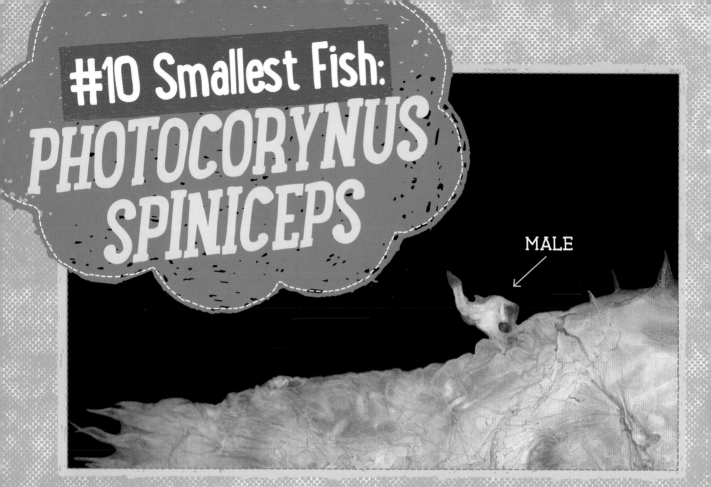

#10 Smallest Fish: PHOTOCORYNUS SPINICEPS

MALE

The male *Photocorynus spiniceps*, a type of anglerfish, is the smallest fish in the world. That's a big name for such a small fish. It's the scientific name. This fish doesn't have a common name yet. How small is it? The male of this **species** is much smaller than the female. It is less than 0.25 inches (6 mm) long.

Males of this species attach themselves to the backs of females and will spend their entire lives there. The females take care of swimming and eating for both of them. The female can weigh thousands of times more than the male.

Like other anglerfish, the female *Photocorynus spiniceps* draws in prey using a **bioluminescent**, or glow-in-the-dark, lure that dangles above its head in the dark ocean. Once prey gets close to the lure, these fish spring forward and swallow the prey whole. The prey can sometimes be as big as the anglerfish.

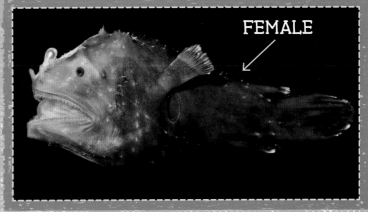

FEMALE

SAME SIZE AS . . .

The male *Photocorynus spiniceps* is smaller than a grain of rice.

Photocorynus spiniceps was discovered in the Philippines.

PHOTOCORYNUS SPINICEPS CLOSE-UP

MALE

EYES
Males have large eyes to help them see a female's lure.

MOUTH
The male bites down on the female's back to attach himself to her.

NOSTRILS
Males have a good sense of smell to help them find females.

The male and female of this species work together to survive.

FACT
The male *Photocorynus spiniceps* is considered a **parasite** because it cannot live on its own. It must latch onto a female to survive.

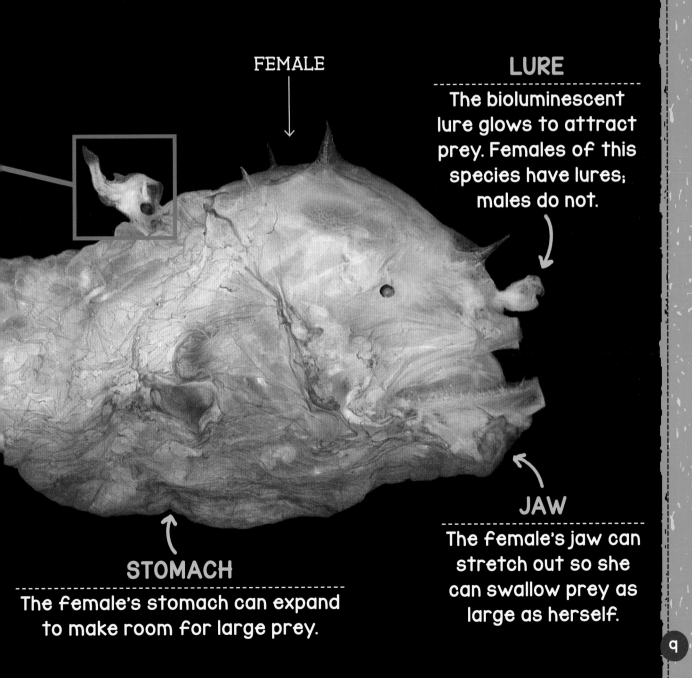

FEMALE

LURE
The bioluminescent lure glows to attract prey. Females of this species have lures; males do not.

JAW
The female's jaw can stretch out so she can swallow prey as large as herself.

STOMACH
The female's stomach can expand to make room for large prey.

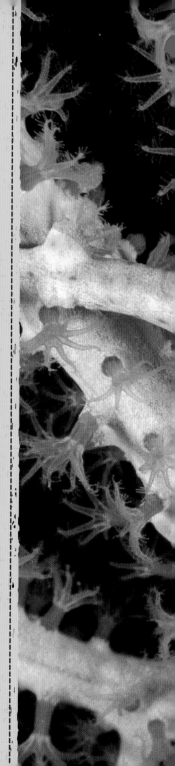

#9
DENISE'S PYGMY SEAHORSE

Denise's pygmy seahorse is one of the smallest seahorses in the world. They are less than 0.8 inches (2 cm) when fully grown. But their size isn't what makes them hard to spot. These seahorses use a special **camouflage** to stay hidden. They can change their color to match their surroundings. Shortly after birth, these seahorses look for a specific type of sea fan coral. Once there, they wrap their tails around the fan and blend in. They will spend their entire lives in this new home.

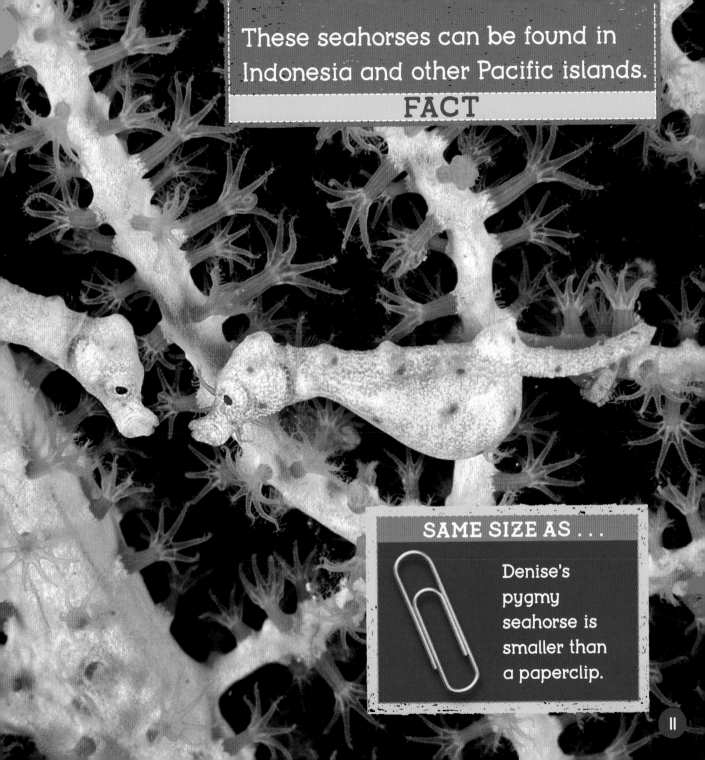

These seahorses can be found in Indonesia and other Pacific islands.

FACT

SAME SIZE AS . . .

Denise's pygmy seahorse is smaller than a paperclip.

This fish is found on reefs in the Indian and Pacific Oceans.

FACT

SAME SIZE AS . . .

A bumphead parrotfish is as tall as a hockey net.

BUMPHEAD PARROTFISH

Bumphead parrotfish can grow up to 4 feet (1 m) long and weigh 165 pounds (75 kg). How do they get so big and heavy? They eat **algae**, which grows inside coral. To reach the algae, parrotfish use their beak-like teeth to rip off small chunks of coral. The eaten coral passes through their body as sand. A single bumphead may produce several hundred pounds of sand a year!

#7
LION'S MANE JELLYFISH

Lion's mane jellyfish are one of the largest jellyfish species in the world. The bell, or top part of their body, can reach lengths of up to 6.5 feet (2 m). Their "manes" are made up of long, hair-like **tentacles**. As many as 1,200 tentacles can hang from the undersides of their bells. These tentacles can be quite long. The longest lion's mane tentacles ever recorded measured 120 feet (37 m) long!

Most lion's mane jellyfish live in the Arctic, northern Atlantic, and northern Pacific Oceans.

FACT

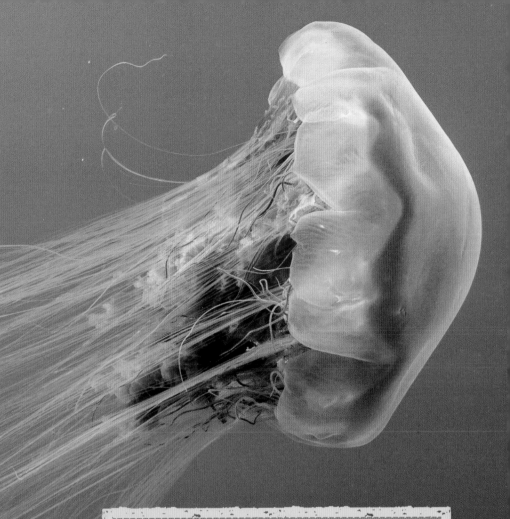

SAME SIZE AS . . .

An average lion's mane jellyfish body is as long as a small couch.

SAME SIZE AS . . .

A goliath grouper is about as long as an adult-size surfboard.

#6
GOLIATH GROUPER

Goliath groupers live up to their name. They can grow to lengths of 8.2 feet (2 m) and weigh up to 800 pounds (363 kg). This species catches prey by blending in with its surroundings. When it spots something to eat, it rushes forward to snap the prey up with sharpened rows of teeth. But the goliath doesn't chew its food! Most food gets swallowed whole. Goliaths eat shrimps, crabs, lobsters, parrotfish, stingrays, young sea turtles, and octopuses.

FACT

This fish can be found in the Atlantic Ocean.

17

MEKONG GIANT CATFISH

Mekong giant catfish are the largest freshwater fish in the world. They can grow up to 9.8 feet (3 m) long and weigh up to 646 pounds (293 kg). These fish have one of the fastest growth rates of any fish. It may take only six years for one to reach 440 pounds (200 kg). As babies, they feed on **zooplankton**. After about a year, they become **herbivores** and eat plants and algae.

SAME SIZE AS . . .

The Mekong giant catfish is about as long as a stepladder.

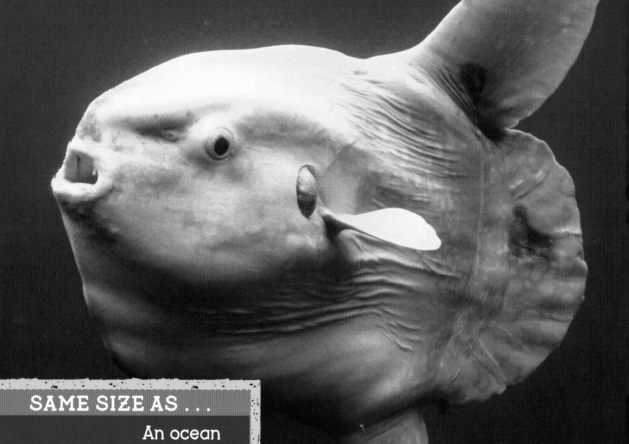

SAME SIZE AS . . .

An ocean sunfish is nearly the size of a Volkswagen Beetle.

OCEAN SUNFISH

Ocean sunfish are huge, flat, and circular. They top out at around 5,000 pounds (2,268 kg) and can be longer than 10 feet (3 m). Inside an ocean sunfish's tiny mouth are two pairs of hard teeth plates. Ocean sunfish mainly eat jellyfish. They don't break their food down into smaller pieces by chewing it. Instead, they suck jellies in and out of their mouths until the jellies become a mushy pulp that they can swallow.

FACT

This fish can be found in the Atlantic, Pacific, and Indian Oceans.

GIANT OCEANIC MANTA RAY

Giant oceanic manta rays are the largest rays in the world. Their bodies can grow to be 16 feet (5 meters) long, and many giant rays have a wingspan up to 29 feet (9 m) wide. They can weigh up to 5,300 pounds (2,404 kg). These rays are **filter feeders**. They eat large quantities of zooplankton. To feed, they hold the fins on their faces in an O shape and open their mouths wide. Water and tiny prey flow into their mouths. Manta rays do something called barrel rolling to feed, which is like doing somersaults over and over again.

These manta rays can live up to 40 years.

FACT

SAME SIZE AS . . .

A giant oceanic manta ray's body is as long as a canoe.

BASKING SHARK

Basking sharks are the world's second-largest fish. They can measure 26 feet (8 m) long and weigh 10,000 pounds (5 metric tons). Their gaping mouths are filled with about 1,500 tiny hooked teeth. One of only three filter-feeding shark species, basking sharks sift through tons of seawater an hour to find tiny zooplankton to eat. As the water strains through their mouths, the zooplankton get trapped in their mouths and swallowed.

FACT

These sharks can be found in the Atlantic and Pacific Oceans.

#1 Biggest Fish: WHALE SHARK

Slow-moving, filter-feeding whale sharks are the largest fish in the sea. They can reach lengths of 40 feet (12 m) or more. They can weigh up to 41,000 pounds (19 metric tons). This large animal has a taste for small prey.

SAME SIZE AS . . .

A whale shark can be longer than a school bus.

Whale sharks scoop up tiny zooplankton and fish eggs with their gaping mouths while swimming close to the water's surface. More than 1,585 gallons (6,000 l) of water an hour pass through their gills.

The biggest fish in the world **migrate** to eat. Whale sharks take their time when they travel, moving barely 2.5 miles per hour (4 kph). They may be slow, but they travel far. Each year, whale sharks may travel up to 2,485 miles (4,000 km) during their migration.

FACT These fish have no known natural predators.

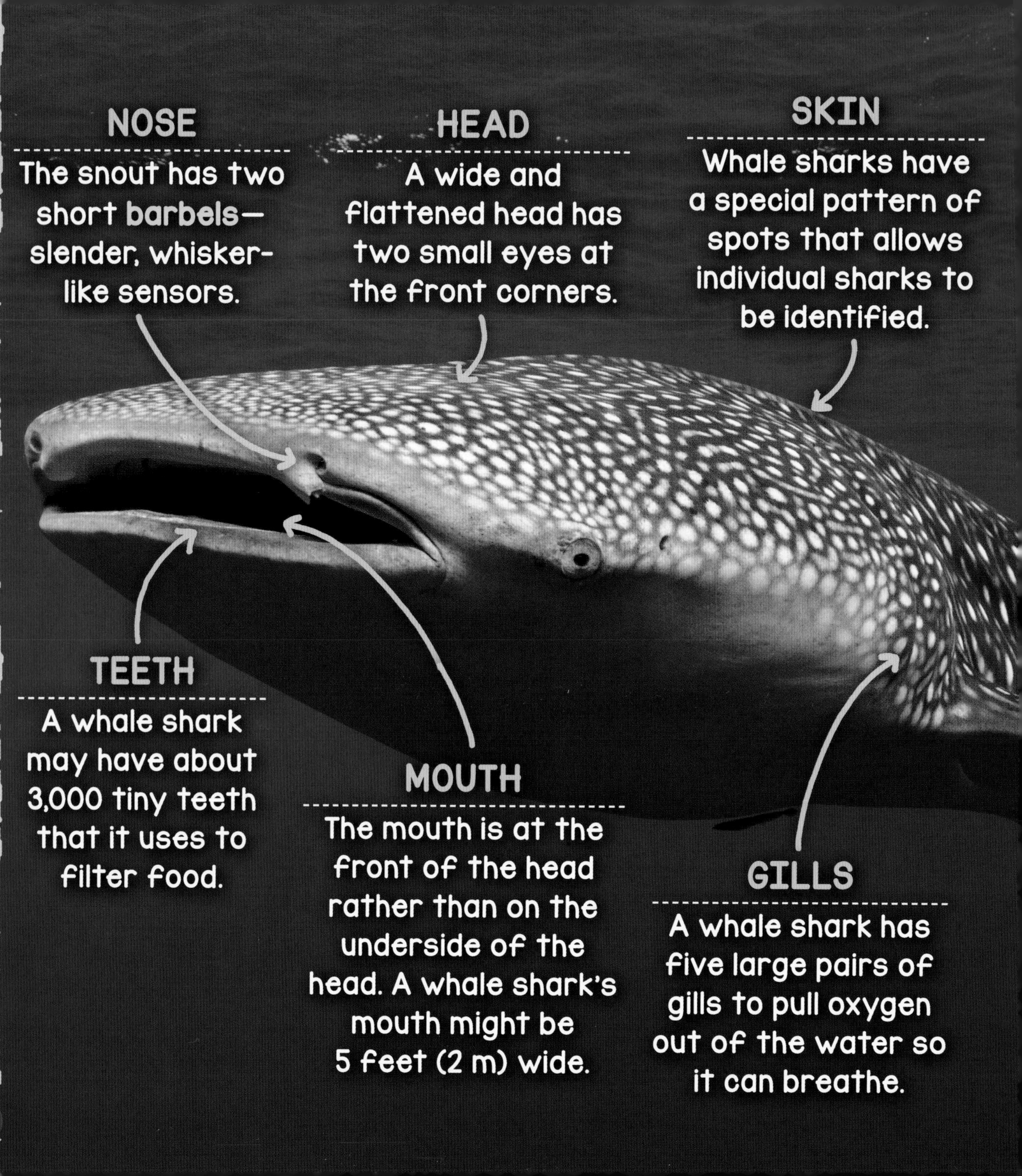

NOSE

The snout has two short **barbels**— slender, whisker-like sensors.

HEAD

A wide and flattened head has two small eyes at the front corners.

SKIN

Whale sharks have a special pattern of spots that allows individual sharks to be identified.

TEETH

A whale shark may have about 3,000 tiny teeth that it uses to filter food.

MOUTH

The mouth is at the front of the head rather than on the underside of the head. A whale shark's mouth might be 5 feet (2 m) wide.

GILLS

A whale shark has five large pairs of gills to pull oxygen out of the water so it can breathe.

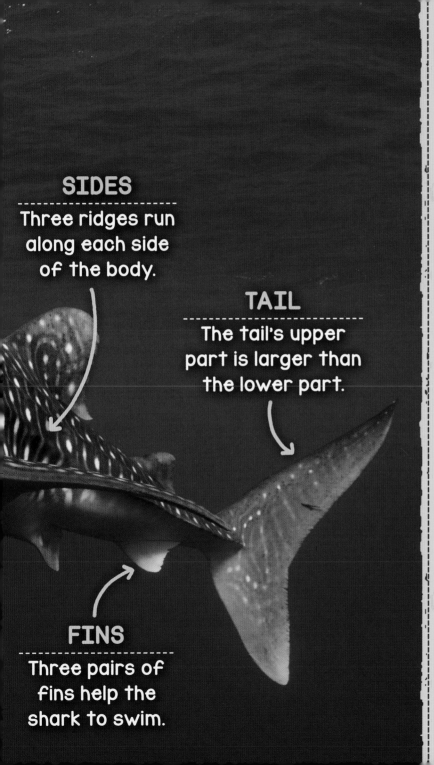

SIDES

Three ridges run along each side of the body.

TAIL

The tail's upper part is larger than the lower part.

FINS

Three pairs of fins help the shark to swim.

WHALE SHARK CLOSE-UP

The whale shark is not just the world's biggest fish. It is also one of the world's longest-living fish. It can live more than 100 years.

FISH BIG AND SMALL

Fish come in all shapes and sizes. Now you know why it matters to be a big or small fish. There are more than 36,000 species of fish in the world—too many to cover in this book! Make it your mission to learn even more about these amazing animals, both big and small.

GLOSSARY

algae (AL-jee) small plants without roots or stems that grow mainly in the water

barbels (BAHR-buhlz) slender, whisker-like sensory organs

bioluminescent (beye-oh-loo-muh-NE-suhnt) a light produced by a living organism

camouflage (KAM-uh-flahzh) to disguise something so that it blends in with its surroundings

cold-blooded (KOHLD bluhd-id) having a body temperature that changes according to the temperature of the surroundings, like reptiles or fish

filter feeder (FIL-ter fee-der) an aquatic animal that feeds on particles or small organisms strained out of water by circulating them through its system

fin a part on the body of a fish shaped like a flap that is used for moving and steering through the water

fish a cold-blooded animal that lives in water and has scales, fins, and gills

gills (gilz) the pair of organs near a fish's mouth through which it breathes by extracting oxygen from the water

herbivore (HUR-buh-vor) an animal that eats plants

migrate (MYE-grate) to go from one place to another

parasite (PAR-uh-site) an animal that lives on or inside another animal

predator (PRED-uh-tur) an animal that lives by hunting other animals for food

prey (pray) an animal that is hunted by another animal for food

scales (skalez) thin, overlapping waterproof plates that cover a fish's skin

species (SPEE-sheez) a group of similar organisms that are able to reproduce

tentacle (TEN-tuh-kuhl) one of the long, flexible limbs of some animals, such as octopus or squid

zooplankton (zoh-uh-PLANGK-tuhn) a collection of small, passively floating, drifting, or somewhat mobile organisms occurring in a body of water

INDEX

Page numbers in **bold** indicate images.

ABOUT THE AUTHOR

Brenna Maloney is the author of more than a dozen books. She lives and works in Washington, DC, with her husband and two sons. She wishes she had more pages to tell you about fish. She also wishes she had the brainpower of a manta ray, the toughness of an amberjack, and the beauty of a mandarinfish. (Look them up!)

BIG NIGHT

BIG N!GHT

Dinners, Parties
& Dinner Parties

KATHERINE LEWIN

UNION
SQUARE
& CO.

NEW YORK

For Alex, my favorite cohost

U

UNION
SQUARE
& CO.

NEW YORK

Text © 2024 Katherine Lewin
Photographs © 2024 Emma Fishman
Photographs on pages 8 and 10 © 2024 Julie Goldstone

Sabzi Polo, page 46, excerpted from *Maman and Me*,
by Roya Shariat and Gita Sadeh, PA Press, 2023

ISBN 978-1-4549-5213-8 (hardcover)
ISBN 978-1-4549-5214-5 (e-book)

For information about custom editions,
special sales, and premium purchases, please
contact specialsales@unionsquareandco.com.

Printed in India

2 4 6 8 10 9 7 5 3 1

unionsquareandco.com

Editor: Amanda Englander
Designer and Illustrator: Mia Johnson
Photographer: Emma Fishman
Food Stylist: Monica Pierini
Prop Stylist: Stephanie de Luca
Art Director: Renée Bollier
Project Editor: Ivy McFadden
Production Manager: Kevin Iwano
Copy Editor: Terry Deal

CONTENTS

I'M SO GLAD YOU'RE HERE!

Hi, hello, nice to meet you! Thank you so much for coming over and opening this book. My name is Katherine, and I own Big Night—a dinner, party, and dinner party essentials shop in New York City.

The idea for Big Night came to me over the holidays in 2020, during the COVID-19 pandemic lockdown. I had spent the prior five years leading a team of restaurant writers and editors—a wildly amazing job that required nonstop thinking about (and eating) some of the best food in the country. Then 2020 happened, and my job went from experiencing restaurants in their dining rooms to experiencing them via takeout, primarily eaten in the front seat of my car.

When I wasn't writing about the way restaurants were innovating, pivoting, and surviving in that impossible year, **I was cooking.** More than I had ever cooked. And I was watching my friends and family from afar, doing the same—The sourdough! The windowsill scallions! The banana bread! The 100-step culinary journeys!—some of them spending real time in their kitchens for the first time ever. And yet, we were all mostly cooking for just ourselves and a roommate, a partner, or a small pandemic pod.

I started thinking about how much **I *missed* dinner parties.** Just like I missed sitting down at a restaurant bar and enjoying a solo martini. Or boarding a plane to Rome. Or walking around the Met, totally unconcerned with how many feet were between me and strangers.

I missed cooking for people. Gathering my friends in my home. The long, winding nights spent together, eating, drinking, laughing, and inevitably dancing on the couch to the Chicks and/or Shania Twain. Did I even *remember* how to host? The idea of cooking for a group of people already seemed foreign at that time. *What do I put on the table? How many dishes is the right number of dishes? Do I own any serving plates?*

After so many months spent on our screens, I knew I couldn't be the only one who was craving community, fantasizing about the dinner parties we'd have and the togetherness we'd feel once it was safe enough to gather. *What if there was a place*, I thought, *that helped remind us all how to do exactly that?* A place that made hosting easy and fun, where you could get everything you might need to **make a night at home feel more special:** A couple of great cheeses—and cheese knives while you were at it. A new-to-you ingredient for a project recipe—plus a beautiful platter deserving of your masterpiece. Colorful wine glasses to dress up delivery pizza.

I dreamed about that kind of place. And when I stumbled upon a tiny-but-mighty 240-square-foot jewel box with floor-to-ceiling windows in Greenpoint, Brooklyn

(my own neighborhood!), I knew I needed to create that place: a one-stop shop for making any night—whether you're hosting a dinner party for eight or hosting yourself for a solo night in front of your TV—a Big Night.

In early June 2021, about six months after that first inkling of an idea, I signed the lease. The next day, I gave my notice at work, leaving one dream job for another. I opened Big Night just over two months later, in August 2021, and quickly learned that **I wasn't the only one who had missed dinner parties.** A year and a half later, I opened our second location, in the West Village of Manhattan. Today, I am so lucky to meet new friends in our shops every day, who love to eat, cook, and host—people who have visited us from all around the country, and the world. **They (and now you!) are my kind of people.**

This book bottles everything I know about hosting at home, around food. I've learned so much from our incredible vendors (who make the very best ingredients in the world), from our customers (who constantly inspire me in their own kitchens and dining rooms), and from my own hosting experiences—both wonderful and successful, and less successful (but still wonderful). And now, I'm so excited to share all of it with you.

anyone can be a host—and any night can be a big night

Before we go any further, I need to tell you something. I may own a dinner party shop, and I might have written a book about dinner parties, but here is my not-so-secret secret: **I'm an introvert.**

As much as I love being with friends, having them in my home, cooking for them, catching up with them, I am the first to admit that hosting can be draining (to say the least). My solution: the **Social Nap**—the inevitable moment of any dinner party, during which, at some point in the evening, I escape to my bedroom, lie down in the dark, and take a few minutes to recharge by myself. I have zero shame about my Social Naps. If I disappear, my friends all know where I've gone. I come back about 10 minutes later, refreshed and ready for the rest of the night.

I don't tell you this just to officially endorse the Social Nap. I tell you this to say, confidently, that **anyone can be a host.** You don't need to be an extrovert. You don't need to be a great cook. You don't even need to have a dining room table.

The first and most important step of hosting is simply **making the decision to gather at home.** Feeding people in your own space creates a special feeling that's hard to replicate. Whether you're serving a meal you spent all day making, or takeout you picked up and put in your prettiest

bowls—that's a Big Night. Which brings me to my next point: **Any night can be a Big Night.**

A Big Night is any night you choose to make a little more special. It's any night you find something to celebrate. It's any night you bring joy, in the form of food, drinks, and/or gathering, to your favorite people. Or even just yourself.

The idea of the "dinner party" often conjures images of perfectly set tables, coursed meals, and stress and anxiety (for the host, and maybe for the guests, too). That is exactly what I *don't* want for my own Big Nights, or for yours. I can never figure out how to orchestrate a meal so that every dish is exactly the right temperature served at exactly the right time. I don't own a single matching set of plates or glasses. **I don't know how to "entertain," and I don't expect you to know how either.** In fact, the moment I became a better host is the moment I realized entertaining is something you do *for* people. Experiencing is something you do *with* them. **A Big Night is about experiencing something together.**

this book = your back-pocket big night guide

Having people over gives you the chance to offer your friends and family something that reflects *you*—your favorite dish, a playlist you created for the evening, the bottle of wine you can't wait to share, not to mention your very own home, filled with all the little quirks that make your space *your* space. Hosting is inherently personal, and no one will do it exactly like you do it. **That's exactly why this book is meant to be a guide—not an instruction manual.** I want to give you tools and inspiration and ideas and my own go-to moves for hosting success, and then I want you to make them yours.

I've always wished I had a **trusty companion for hosting.** Not a cohost, but rather, a resource I could turn to for menu inspiration that wasn't just going to point me in the direction of yet another soup or stew. I mean, soups and stews are great, but why is it so hard to find more recipes specifically for bigger groups? I've written this book to be that kind of companion: a treasure map full of tips for how to turn any night into a Big Night, and to remember why you'd want to do that in the first place.

This book is mapped out like a calendar year—it's split into four chapters, one for each season. There are eighty-five recipes, all of which feed six or more people (with the exception of one date night menu on page 229, which is for just two). I created

every recipe included here with the hope that it will lead you to your next Big Night— whether it's a Party Chicken kind of night (page 27), or a Big Chopped Salad (to Go with Takeout Pizza) kind of night (page 162).

In every season, you'll find three Bigger Nights—all inspired by my own experiences in hosting—where I've compiled recipes into menus intentionally designed to balance your time and energy, and offer a variety of flavors and textures. By no means should you ever feel like you have to make all of the items suggested in a Bigger Night—these are options, not assignments! I would love nothing more than for you to combine the recipes across Bigger Nights, and across this book (as well as other cookbooks), to make your own perfect menus. To help with that, you'll also find dish pairing suggestions alongside many of the recipes.

The least stressful, most successful hosting nights are the ones in which at least a good chunk of the actual cooking work is done in advance. To that end, I've included **Make-Ahead** indicators wherever you can get away with prepping these recipes, either in part or in full, long before anyone walks through your door. You'll find a shorthand summary of which elements can be made in advance, along with more detailed notes in *italic* throughout. Please feel entirely free to skip over all

of that if it's not relevant to you in that moment. What I would implore you to *not* skip is reading the entire recipe before you start to cook. The last thing I want is for you to run into a step that you didn't account for when you decided to make the recipe. I love surprises, but not that kind.

Keep in mind: Serving sizes are estimates. Every body is different, and every meal is different. Something that took me a long time to fully wrap my ahead around: When a recipe says it "Serves 6," that does not necessarily mean 8 or 10 or 12 people wouldn't be perfectly fine to share it—it just depends on what else you're serving. The best way to get better at making these calculations for yourself is simply to keep hosting and learning how much food people really need. (I personally live in fear that people will leave my house hungry, and always cook too much, but I'm working on it. Plus, leftovers are the best.)

Lastly, I'm excited to introduce you to some **special guests.** I've invited potluck-style contributions from five people I really admire, each of whom has made me a better home cook and host in some way. Besides, this book didn't feel quite like a party without a few friends.

If this book makes you want to make dinner tonight, I'll be thrilled. If it makes you want to host a dinner party tonight, I'll be even happier. But most important: I hope this book reminds you that **you deserve more Big Nights, any and every night of the week.**

DINNER & PARTY ESSENTIALS

the big night pantry

One of the reasons why my job is the absolute greatest is that I have the privilege of tasting, and then sharing, the most delicious ingredients and pantry staples from all over the world. Even better: I get to know the people who make them. I am reminded each day, thanks to these makers, that the best ingredients not only make for the best meals, but they also transform a dish in a matter of seconds.

Rather than telling you about the lemons and garlic and cream you'll find listed in similar sections in a lot of cookbooks (these lists are important—you just don't need yet another one from me!), I want to tell you about the **highly specific brands and goods I personally use most often.** These are the jars, bottles, and boxes I turn to again and again—whether I'm looking for the perfect salad-finisher or a way to zhuzh my takeout leftovers—and how in particular I use them:

EXTRA-VIRGIN OLIVE OIL

Fat Gold, Graza "Sizzle"
● For cooking over any kind of heat; using in sauces and dips

Wonder Valley, EXAU, Oracle
● For a finishing drizzle on salads, pastas, and proteins of all kinds

SALTS & SPICES

Diamond Crystal Kosher Salt
● For **seasoning every single dish I cook, including every recipe in this book, unless otherwise noted.** (If you cook with a different type of salt, these dishes will come out very differently!)

Maldon Salt
● For a final flaky sprinkle on anything that could use a salty crunch (which is most things)

Diaspora Co.
- For black pepper, turmeric, and cinnamon so fresh and fragrant, you'll wonder where they've been all your life

Daphnis and Chloe
- For the best dried thyme and oregano

DRIED BEANS & PASTA

Beans: Rancho Gordo, Primary Beans
Pasta: Benedetto Cavalieri, Monograno Felicetti, and Martelli are worth the price for a special meal; De Cecco is forever my everyday pasta

SAUCES & SUCH

Hot Sauce: Zab's
Soy Sauce: Cabi
Wine Vinegar: Brightland, Camino
Balsamic Vinegar: Sardel
Tahini: Seed & Mill
Gochujang: Queens SF
Miso: Moromi
Canned Tomatoes: Bianco DiNapoli

FINISHING TOUCHES

Onino, KariKari, Boon
- For that hit of spicy, salty, sweet, crunchy flavor only chili crisps can provide

Woon Sea Moss Seasoning
- For a sprinkle of umami on eggs, avocado, rice, and salads that tastes like the ocean

Daphnis and Chloe Smoked Chili Flakes
- For a smoky, spicy finishing note for pastas

Bungkus Bagus Sambal Goreng
- For a fried garlic & chili crunch for scattering over roasted veg or fish

---

snack spread essentials

I'M NOT THE MOST Always-Prepared person—but with these snacky, appy items handy, I feel fully equipped to have people over on a whim, always ready for a Big Night:

- Great Olive Oil: For crusty bread, for drizzling on dinner, or for zhuzhing up store-bought dips

- Tinned Fish: Fishwife, Güeyu Mar, and Minnow are just a few of my favorites

- Chips: Ruffles for everyday, Bonilla a la Vista for special-occasion snacking

- Crackers: I love Unbothered Sourdough Crackers and Mitica Toketti

- Cheese: A firm cheese (aged cheddar, Gouda) and a soft cheese (anything Brie-style, chèvre)

- Charcuterie: A salami or prosciutto or both

- Olives: Castelvetrano, Gordal, or a mixed bunch

- Nostagic Faves: Oreos for sundaes, Ritz for dips, Swedish Fish for anytime (trust me)

the big night kitchen

I learned how to cook—and then, how to host—within the very small bounds of my New York City apartment. I do not have a Nancy Meyers movie–size space to store all of my cookware and tableware in organized bliss. **Nor am I, by any stretch of the word, a minimalist.** The cabinets I do have are stuffed to the brim (sometimes beyond the point of complete closure) with kitchen and dining items that fall into three categories:

PURELY PREP

A chef's knife, whisk, and Microplane are the cookware equivalents of pantry staples like lemons, garlic, and cream: We already know they're essential. So rather than waxing poetic about how a high-quality, sharp chef's knife will change your life, let me share this checklist of the items I reach for most in my own kitchen:

A 7- to 9-inch chef's knife and a paring knife
Large pasta pot
Rimmed baking sheets
Silicone spatula
Wooden spoons
Tongs
Whisk
Microplane
Cheese grater
Oven thermometer and instant-read thermometer

Large fine-mesh sieve and colander
Measuring cups and spoons
Hand mixer (while I love my stand mixer, it takes up an unfair amount of space—currently on top of my refrigerator)

DOUBLE-DUTY: COOK & SERVE

When I'm buying a new piece of cookware, I invest in the kinds of pots and pans that bring as much joy to my kitchen as they do to my dining table. If I have the opportunity to choose between a colorful cutting board and a neutral one, I'll always pick the former. Life is too short, and my kitchen is too small, to deny myself the pleasure of a rainbow cutting board—plus, it's a rainbow cutting board that can also be used as a rainbow serving board. Here are my MVPs for both making and sharing dinner:

Large cutting board: That can also be a cheese board, a snack board, or a serving plate for main dishes
Medium cutting board: For serious chopping that won't take up too much counter space, or a smaller snack plate
Small cutting board: You'll want more than one of these—for bartending, quick garlic mincing, or serving a handful of nuts
12-inch cast-iron skillet: For any searing moments and sometimes even baking, then bringing straight to the table

Medium saucepan: I don't own a small saucepan, and I don't feel like I need one, but I love my medium-size Dansk Købenstyle for all manner of saucy situations

6- to 8-quart Dutch oven: My red Le Creuset is the most-used item in my kitchen. I use it for pasta sauce, soups, and braises, cooked on the stovetop or in the oven and often served straight from the pot.

Mixing bowls: Small, medium, large, and all the way up to XXXL (the secret to great salads is the biggest bowl you can find)

2- to 3-quart baking dish: For roasted fish, baked dips, and desserts

Lots of little bowls: For mise en place, for olive pits, for condiment serving stations, for when you forget to take your jewelry off before you start cooking and you want to put it somewhere safe.

TABLE ONLY

A platter really is so much more than a platter. At least, that's how I feel. It's not just a vehicle for "plating." It's the final, crowning step of any dish that you put your time, energy, and care into. **A platter is a thing that says to your guests, "See, this is how much I love you."** It's a thing that you bring to your table over and over again. And it's a thing that then shows up in your photos and memories. It's a thing that, hopefully, brings happiness every time you use it.

Which is exactly why I would advise: Don't just buy a platter because you think you "need" a platter. Not all your platters have to look the same or like everyone else's. Wait until you find the platter that speaks to you. Yes, I do believe that objects speak to us; you just have to listen! And that goes for every other piece for your dining table.

The more serving-vessel options you have, the better (as I said, I am not a minimalist). Anything—and I mean anything—can look more delicious in the right platter or bowl. These are the serving pieces I own in multiples:

Snack dishes: For olives, dips, nuts

Blates: Bowl-plates, the most useful dinnerware that can double as serving vessels

Medium serving platters: For vegetable sides; I like an oval shape

Large serving platters: Ideally with a lip for catching sauces and juices

XL serving bowls: For salads or piles of chips

XL serving platter: Like, big enough for a turkey or a truly grand snack spread

Dessert dishes: Ideally clear, so you can see your ice cream through them

the big night bar

It's one thing to make a perfect Negroni or spritz or martini for yourself; it's quite another to be your dinner party's bartender, hurriedly making different drinks for different people, not actually getting to talk to anyone, and stressing out that you're leaving dinner in the oven for too long because you're in the middle of making, shaking, stirring, and/or pouring.

Personally, **I don't believe anyone should be bartending all night**—not even the host(s). I pick one or two house cocktails for the night, plus a nonalcoholic option: drinks that are either simple to make, possible to batch, or, even better, beverages guests can put together themselves. In the same way I think about the flow of food throughout the evening, I like to have a plan in mind for drinks, too. For the purposes of menu planning, I divide cocktails into four main categories:

- **Best Before Dinner (BBD):** You can pretty much get away with serving any cocktail before dinner. I typically go for drinks on the lighter side of things (a Big Night is a marathon, not a sprint): spritzes and other cocktails with soda water or sparkling wine as a main component, which helps keep the ABV a little lower.

  - Any-Amaro Spritz (page 143)
  - Negroni Sbagliatos (page 31)
  - Improved Mezcal Palomas (page 58)

- **Best After Dinner (BAD):** Anything amaro-based is fantastic after dinner, as is a boozier drink like a Manhattan or an old-fashioned. Oh, and espresso martinis are, unequivocally, a Best After Dinner drink. I will not be accepting any other opinions on that matter at this time.

  - Batched Hanky-Pankys (page 261)

- **Best with Dinner (BWD):** Ah, the elusive BWD. I usually prefer to serve wine at dinner—in large part because everyone can pour themselves. It can also be tricky to find a cocktail that pairs nicely with food, but I tend to find that simpler, spirit-forward drinks can do the trick.

  - Classic Negroni (page 29)
  - White Mezcal Negroni (page 31)
  - Ideal Whiskey Highball (page 172)
  - Wine (see page 213)

- **Zero-Proof (NA):** So many wonderful nonalcoholic beverages have sprung up in the last few years. Here are just a few of my favorites:

  - Ghia and Figlia (over ice, with club soda and a twist of lemon or wedge of an orange)
  - St. Agrestis Phony Negroni
  - For Bitter For Worse
  - Unified Ferments (technically a kombucha, this reads like a natural wine, and deserves a wine glass accordingly)

With crowd control and menu planning out of the way, let's talk about the cocktails themselves. Great cocktails are made with great ingredients—and I'm not just talking about the booze. Ice matters. The type of club soda you top off the drink with matters. Glassware matters. That last one just might be my favorite cocktail topic. Here are the types of cocktail glasses I use most often in my own home, and throughout this book:

- **coupe glass:** for anything served up (without ice), and for sparkling wine
- **rocks glass:** for Negronis of all kinds
- **lots-of-ice glass:** (wine, tumbler, and/or Collins) for spritzes and highballs

three bartending tips

- Make or buy extra ice. More than you think you need.

- Prep your garnishes in advance: Zest long strips of lemon or orange for Negronis or spritzes; cut wedges of grapefruit for Palomas. (My bartender friends like to go an extra step here and cut off the edges of the citrus wedges in straight lines, which I have to admit, really does make them look so much nicer.) While you're at it, feel free to have fun and combine garnishes: I love an olive and a lemon peel in my spritzes.

- If you have space in your freezer, freeze the first round of cocktail glasses. It's a simple step that will make everyone feel like royalty.

hosting rules I live by

1. **You don't have to serve dinner to have a dinner party.** Snacks can be dinner. And if those snacks run out, pizza is always a phone call away.

2. Rather than a hard start time, **give people an arrival window**: "Come over any time after six thirty—we'll eat around eight." This framing accomplishes a lot: it tells people they can expect drinks and snacks before dinner, it gives them flexibility if they need a little more time, and it conveys an actual deadline for them to show up by. All in one clearly communicated sentence!

3. **Anything—even a store-bought thing—looks good in the right bowl.** Do not underestimate the power of a great (or unexpected) serving vessel. Whip out a gravy boat filled with hot fudge for pouring over ice cream, and watch the people freak out.

4. **Right bowl, right size.** If you've ever wondered why your dip, or salad, or cheese plate looks a little sad, consider: Does it fill out the vessel you put it in? Fit is important; don't go too big. **Your food should be wearing its outfit, not the other way around.**

5. **Anything can be zhuzhed.** Good olive oil, flaky salt, freshly cracked black pepper, a sprinkling of chile flakes, a fistful of fresh chopped herbs, or a scoop of chili crisp can turn a meh dish into a much better one—or even breathe new life into leftovers.

6. Sometimes—most times— **the best appetizers are the simplest ones**: A hunk of great cheese, a bowl of olives, and a little pile of salted nuts make for a perfect starting spread.

7. **If you're serving cheese, take it out of the fridge at least half an hour beforehand.** You want to be able to taste the cheese, not your fridge. These thirty minutes can be the difference between a just-pretty cheese plate and an utterly delicious one.

8. If at any point you feel overwhelmed or stressed by the menu you've planned to serve, **delete a dish**. No one will know, and you won't be cranky and stressed.

9. **CAYG: Clean as You Go.** It's a lifestyle—one that is, for me, sometimes more of an aspiration, but still. The more I tidy and wash as I cook, the happier I am at the end of the night not to be staring at a Mount Everest of mess.

10. **If you can turn something into a bar, do it.** A Spritz Bar, a Deviled Eggs Bar, a BLT Bar, a Bloody Mary Bar—people love a bar, they love options, and they love to customize. Everyone wants to be their own special snowflake! And hosting is so much easier when everyone becomes the master of their own destiny.

11. **Don't forget dessert.** It can be as simple as fancy chocolate bars broken up and shared right on the tablecloth. It can be as nostalgic as Klondike Bars or Chipwiches or Otter Pops pulled from the freezer. It can be as easy as pastries outsourced from your favorite bakery. It can be as fancy as a three-layer cake you made from scratch—but if you're going this route . . .

12. **Don't stress dessert.** Only make it yourself if you really want to, and do it before the party has started. Your maximum dessert effort should be scooping it into glasses, reheating it in the oven, or making fresh whipped cream—in fact, simply add some sliced berries tossed with a little sugar, and that right there is a perfect dessert.

13. Accept help with the dishes—but only at the very end of the night.

PARTY CHICKEN WITH FETA & FENNEL

Serves 6 to 8, easily scaled up

2 large or 3 small fennel bulbs, with stalks and fronds (about 1½ pounds total)

3½ to 4½ pounds bone-in, skin-on chicken parts, such as thighs, legs, and/or breasts (breasts halved crosswise, if using)

Kosher salt

½ cup extra-virgin olive oil, plus more as needed

2 (15.5-ounce) cans cannellini beans or chickpeas, drained, rinsed, and patted dry

8 scallions, cut into 1-inch pieces

2 tablespoons za'atar

3 teaspoons mild chile flakes, such as Aleppo pepper or gochugaru, plus more for serving

Freshly ground black pepper

2½ cups pitted Castelvetrano olives

1 (6- to 8-ounce) block feta cheese

2 cups fresh or frozen English peas (about 10 ounces)

1 lemon

Fresh mint leaves, for serving

Crusty bread or cooked rice, for serving

Everything we've ever been told about roast chicken is a lie. Okay, maybe *lie* is a strong word—but it's a myth, for sure. "Perfect for entertaining," they say. "So simple and easy to make," they say. "You may even have leftovers," they say. No, no, and definitely not. For starters: One chicken is simply too small for a group. No matter how elegantly roasted, perfectly crispy-skinned, and positively juicy that bird is, a roast chicken can feed four people *max*—and that's if everyone in the group has compatible chicken desires (what happens if three out of four only like dark meat?). If you're cooking for five or more humans, you really need two chickens. And I, for one, do not have it in me to make two chickens.

The solution? Chicken pieces—not a whole chicken—which can easily be scaled and selected for your group's size and meat preferences, and which cook on sheet pans for weeknight ease. But don't let those sheet pans fool you. This chicken is a party, thanks to the mix-and-mingling of white beans for richness, roasted fennel for earthy crunch, olives for can't-stop-eating-it brininess, peas for sweetness, and feta baked in the oven so it's ready to slather all over the crusty bread you should definitely serve with it.

Position racks in the upper and lower thirds of the oven and preheat to 425°F.

Trim off the bottoms of the fennel bulbs. Slice the stalks from the bulbs, then separate the fronds from the stalks. Halve the bulbs lengthwise, then thinly slice the halves. Thinly slice the stalks. Divide the sliced bulbs and stalks evenly between two baking sheets. Reserve the fronds.

Pat the chicken very dry with paper towels (the drier the chicken, the crispier the skin), season on all sides with salt, and divide evenly between the baking sheets. Drizzle the chicken and fennel with about 1 tablespoon of the olive oil per baking sheet, tossing to coat. Bake for 10 minutes.

After 10 minutes, reduce the oven temperature to 350°F and remove the baking sheets from the oven. Divide the beans and scallions evenly between the baking sheets. Shower each baking sheet with 3 tablespoons of olive oil, 1 tablespoon of za'atar, 1 teaspoon of salt, 1½ teaspoons of chile flakes, and a few grinds of pepper. Carefully toss the vegetables, chicken, and seasoning together, then place the chicken on top of the vegetables before returning to the oven. Roast, rotating the baking sheets between racks and from front to back halfway through cooking, until the chicken is mostly cooked through, 25 to 30 minutes.

While the chicken and vegetables roast, use the side of your knife to crush the olives. Cut the feta into 8 pieces.

Remove the baking sheets from the oven. Divide the peas, olives, and feta between the baking sheets, carefully tossing them with the rendered fat in the pans. Return the baking sheets to the oven, rotating their positions again, and roast for another 15 to 20 minutes, until the peas are tender and the chicken is cooked through.

Turn your oven to broil. Move one of the oven racks to the top rack, closest to the broiler. Remove one baking sheet from the oven and place the other under the broiler. Broil until the chicken skin is golden brown and crispy, and the feta is beginning to take on color at the edges, 2 to 4 minutes, checking every minute or so to make sure nothing burns. Remove from the oven and repeat with the other baking

sheet. When the second pan is finished, let both rest, uncovered, for 5 minutes.

Meanwhile, roughly chop about ½ cup of the reserved fennel fronds. Slice the lemon in half lengthwise and cut one half into wedges. Transfer the chicken to a serving plate, on top of the vegetables and feta. Squeeze the remaining lemon half over the dish, then top with the mint and the chopped fennel fronds. Serve with the lemon wedges and bread on the side.

NOTE: When buying your chicken, don't forget to factor in leftovers for the next day—simply toss pulled meat with a little mayo and lemon or chili crisp, and you have a chicken salad lunch party.

PAIR WITH: *Fluffy Sheet Pan Focaccia (page 220) or Fast Yogurt Flatbreads (page 74), Sabzi Polo with Tahdig (page 46), A Super Fresh Yogurt Side (page 51)*

NEGRONI 3 WAYS

Makes 1 of each

· ·

CLASSIC NEGRONI

1¼ ounces gin

1 ounce sweet vermouth (I like Cocchi Torino)

1 ounce Campari

Small pinch of salt

Large ice cube, for serving (see Note)

1 wide strip of orange peel, for serving

1 rosemary sprig, for serving (optional)

NOTE: If all you have is normal-size ice machine cubes, that's fine— you'll still have a delicious cocktail! However, those cubes will melt faster than a large cube would, which means you'll either get a little extra dilution or you have an excuse to drink faster.

This is my house Negroni—the cocktail I drink and serve most often. The classic spec calls for equal parts gin, sweet vermouth, and Campari, but I find that mix a little too sweet and syrupy—the Campari is a LOT, and it tends to drag down the freshness and cleanness of the cocktail. So I up the gin a touch. This is a delicate balance, and one of personal taste—too much gin (say, 1½ ounces, which some people recommend) reads too "hot" and boozy for me, but 1¼ ounces strikes the perfect balance. As for the gin itself, my go-tos are Tanqueray, which is classic, and Bombay Sapphire, which is a bit softer—but really, almost any bottle will be good.

If you have time/remember, slip a rocks glass or two in the freezer.

In a mixing glass, combine the gin, vermouth, and Campari. Add the salt and a good amount of ice and stir until very cold, 20 to 30 seconds.

Place a large serving ice cube in a (hopefully) chilled rocks glass and strain the drink over the top.

Express the orange peel over the drink, rub it around the rim of the glass, and drop it into the drink to garnish. If you have one handy, slap a rosemary sprig (to make the leaves more aromatic) and drop it in as a second garnish.

→

NEGRONI SBAGLIATO

MEZCAL WHITE NEGRONI

CLASSIC NEGRONI

NEGRONI SBAGLIATO

This slightly lower-ABV, refreshing Negroni alternative is lovely pretty much any time of day. I also play fast and loose with the traditionally prescribed proportions here—I go big on the Prosecco, which makes a drink that reads as much like a spritz as it does like its more serious namesake.

1½ ounces sweet vermouth
(I like Cocchi Torino)

1 ounce Campari

4 to 5 ounces good Prosecco

1 wide strip of lemon or orange peel, and/or a Castelvetrano olive, for serving

In a mixing glass, combine the vermouth and Campari. Add a good amount of ice and stir until very cold, 20 to 30 seconds.

Fill a chilled wine glass or rocks glass with ice and strain the drink into it. Top with Prosecco and stir once to combine. Express the lemon peel over the drink, rub it around the rim of the glass, and drop it into the drink, along with an olive, if desired, to garnish.

MEZCAL WHITE NEGRONI

White Negronis taste like Negronis that took off their business clothes and went for a walk barefoot on the beach. They're lighter, brighter, and sunnier. And when you sub mezcal for gin, this drink tastes even more like a vacation. I also find it makes for a great pairing with many different types of dishes.

1¼ ounces mezcal

1 ounce Cocchi Americano (or another fortified white wine, such as Lillet Blanc or white vermouth)

1 ounce Suze

Small pinch of salt

Large ice cube, for serving (see Note, page 29)

1 wide strip of lemon peel, for garnish

In a mixing glass, combine the mezcal, Cocchi, and Suze. Add the salt and a good amount of ice and stir until very cold, 20 to 30 seconds.

Place a large serving ice cube in a (hopefully) chilled rocks glass and strain the drink over the top. Express the lemon peel over the drink, rub it around the rim of the glass, and drop it into the drink to garnish.

the hosting timeline

I AM A PERPETUAL OVERPACKER. *Yes,* I'll think, *I should definitely pack these six pairs of shoes plus eight dresses (plus three sweaters, just in case!) for a 72-hour trip in the summer.* Same goes for time. I like to think that I am also a perpetual optimist—or maybe, just incredibly stubborn—about how much I think I can get done in any allotment of minutes or hours. *Oh, I have an hour between these two meetings? Let me go to the grocery store, and also get in a pedicure. I will pack! It! All! In!* You can guess how that always turns out.

When it comes to cooking, once I get an idea in my head for what I want to make, no matter how limited my prep time is, the original idea is almost immediately followed by about ten "ands." *Okay, a fun steakhouse-ish dinner tonight . . . so, steaks. And creamy spinach. And mashed potatoes. And . . . onion rings?!* (Having made this exact meal, I can tell you: Do not attempt the onion rings. Or, if you're attempting the onion rings, then that's all you're making for dinner, and honestly, everyone will love it.)

This has all been a very long-winded way of saying: **I am always running late.**

Maybe this sentiment deeply resonates with you. Maybe just reading this made your punctual skin crawl. Whichever camp you fall into, may I suggest: the Hosting Timeline. When I'm hosting, I want—no, need—a timeline. And I think you should have one, too.

What's a Hosting Timeline? Think of it as your personal itinerary for everything that needs to happen on a day when you are hosting. For me, there's an emphasis on *personal*: I like to include things like what time I'm waking up, cleaning my apartment and/or self, and when I'm doing any other activities or errands. While you may not require as much self-supervision, at the very least, your Hosting Timeline should include every major step needed to execute whatever it is you're hosting—not the individual steps of a recipe, but the bigger milestones: when the cheese should come out of the refrigerator to come to room temperature; when which dish should go in the oven; and a list of what can be premade the morning of, or even earlier.

You know how they say that when you're looking into the face of a really big project and don't know where to start, you should break that project up into smaller chunks? Think of this as the hosting equivalent.

Timelines will vary in length and specificity depending on the event. My Thanksgiving timeline starts on the Monday of Thanksgiving week, when I'm picking up groceries, while a weeknight dinner timeline might start the minute I get off work that day and include only two major milestones:

1. Place takeout order.
2. Go to wine store.

Another benefit: The timeline keeps us "optimists" honest. Once you start to plot out your step-by-step plan, you might realize that fourth dish you were trying to squeeze into this meal isn't going to happen. Cut an entire dish out of the menu! No one will know what's missing, and you'll be happier you made that decision

in advance—before you wasted time, energy, and money.

The Hosting Timeline also means you're not having to do mental math once your guests arrive. Your job, just like your guests' job, is to actually enjoy yourself. And you know what's not particularly enjoyable? Thinking about when the chicken needs to come out of the oven, while feeling pangs of anxiety that maybe it's already overcooked, while also realizing the greens still need to go in, while trying to carry on a conversation with someone.

Now, does this mean your Big Nights will always go according to schedule? Of course not. But course-correcting on the fly, pivoting, or simply telling people the chicken needs a bit (or a lot) more time is always easier when it's not coming from a place of panic but, instead, a place of relative preparedness.

CRUNCHY, ROASTY GLITTER

Makes about 2 cups

1 cup raw pistachios, almonds, or walnuts

⅓ cup shelled raw sunflower seeds or pumpkin seeds

3 tablespoons sesame seeds

2 teaspoons cumin seeds or coriander seeds

1 teaspoon fennel seeds

1 teaspoon flaky sea salt, plus more to taste

1 tablespoon nutritional yeast (optional)

MAKE-AHEAD: *Up to 2 months*

SPRINKLE ON: *Crunchy, Creamy Buttermilk Slaw (page 103), Veg & Dip Spread of Dreams (page 38), Herby Double Summer Bean Salad (page 110), Valentine Wedge (page 234), Saucy Sesame Spinach (page 173)*

When sprinkled on pretty much anything, Crunchy, Roasty Glitter has the tendency to elicit responses like OH MY GOD WHAT IS THIS? and WHY CAN'T I STOP EATING IT? I live for this kind of back-pocket, never-fail secret that feels like a magic trick. Make a batch and keep it on hand, ready for any and all moments that could use just a little extra sparkle—whether it's your Wednesday lunch al desko, or Saturday evening dinner al fresco. This recipe is meant to be a template for you to explore your ideal combination of nuts, seeds, and spices—and to have fun with variations for different kinds of recipes or moods. The crunchy, roasty, glittery possibilities are truly endless.

Preheat the oven to 350°F.

Spread the nuts in an even layer on one baking sheet and the sunflower seeds on another. Toast until the sunflower seeds are fragrant, about 5 minutes, then transfer them to a medium bowl. Continue toasting the nuts until fragrant, 1 to 2 minutes more. Transfer them to the bowl with the sunflower seeds. Let cool for 5 minutes. Leave the oven on.

Spread the sesame, cumin, and fennel seeds on one of the baking sheets and toast until fragrant, 2 to 3 minutes. Let cool for 5 minutes on the baking sheet.

Working in batches as needed, transfer the toasted nuts and sunflower seeds to a spice grinder or coffee grinder (or a mortar) and grind (or crush with the pestle) until broken down into coarse pieces—like tiny pebbles, not fine sand. It's okay if your mixture includes some bigger and some smaller pieces—they don't all need to be perfectly consistent! Return the mixture to its original bowl.

Transfer the sesame, cumin, and fennel seeds to the spice grinder or mortar. Grind or crush the seeds until just broken open, then add them to the bowl with the nuts.

Mix in the flaky salt, then the nutritional yeast (if using). Taste and add more salt if needed. Let the mixture cool completely, then sprinkle it on anything that could use a little extra something: dips, salads, roasted or grilled vegetables, rice, chicken or fish, even ice cream (just skip the nutritional yeast for that one). *Store in an airtight container or a zip-top bag in the refrigerator for up to 2 months.*

TAHINI-MISO SPECIAL SAUCE

Makes about 2 cups

1 lemon or lime

1 cup tahini

2 to 3 tablespoons sriracha

2 tablespoons red or white
    miso paste

2 tablespoons soy sauce, plus
    more to taste

1 tablespoon honey

¼ to ½ cup kimchi, drained
    and finely chopped
    (optional)

Tahini and miso are both all-star ingredients on their own. Together, they're unstoppable. The combination makes for a dip/spread/sauce that is seriously difficult to stop eating—especially if you add drained and finely chopped kimchi, a little X-factor surprise-and-delight.

Zest the lemon into a medium bowl, then halve and juice it into the bowl. Whisk in the tahini, ⅓ cup water, 2 tablespoons of the sriracha, the miso, soy sauce, and honey. Add more water, 1 tablespoon at a time, as needed, until the mixture is smooth and creamy. Fold in the chopped kimchi (if using), then taste and add more sriracha and/or soy sauce if you like. *Store in an airtight container in the refrigerator for up to 2 days.*

MAKE-AHEAD: *Up to 2 days*

VEG & DIP SPREAD OF DREAMS

Serves 8 to 12

Otherwise known as: dips for dinner. Here is my guide to creating a perfect snacking spread that just happens to be vegetarian. Go big enough with each of the below categories and this could very well be the main eating event—then bring out a killer dessert that everybody will actually have room for.

. . . .

Veg

**PICK 3 OR 4 VEG,
IDEALLY A MIX OF:**

- **Crunchy Veg,** such as cucumbers, snap peas, carrots, fennel, or radishes

- **Leafy and Scoopy Veg,** such as radicchio, Treviso, endive, or Little Gems

- **Cruciferous Veg,** such as broccoli, broccolini, cauliflower

- **Veg That Are Technically Fruit,** such as cherry tomatoes, heirloom tomatoes, or bell peppers

. . . .

Dips

PICK 2 OR 3:

- Tahini-Miso Special Sauce (page 37)

- Green Onion Dip (page 184)

- Ranch-on-Everything Dip (page 90)

- Lighthouse's Hummus (page 154)

- Artichoke Dip (page 186)

- Warm Anchovy Dip (page 270)

- Tonnato (page 138)

. . . .

Seasonings, to Sprinkle on Veg

PICK 2 OR 3:

- Aleppo pepper or gochugaru or Tajín or togarashi

- Furikake or sea moss

- Lemon and olive oil

- Crunchy, Roasty Glitter (page 34)

. . . .

Something Carby

PICK 1 OR 2:

- Fast Yogurt Flatbreads (page 74)

- Big Night Party Mix (page 180)

- Ricotta Toasts for Every Mood (page 44)

- Crusty bread

- Seedy crackers

- Pita chips

AN ALMOST-SPRING THING

SPRING IS NOT JUST A SEASON; it's a state of mind. In New York City, when the very first sunny, clear, 50-degree day arrives, you can feel a palpable euphoria all over the city. People are walking around, giddy, almost (*almost*) smiling at each other, simply because we share in the cautious optimism that we have finally made it out of the never-ending gray-cold of January, February, and March. It's a moment of new beginnings, fresh starts, cleaning out the closet, resetting resolutions. It's a moment that feels worth celebrating.

Growing up in Los Angeles and Austin, two basically always-warm cities, I might not have experienced such seasons in the traditional sense, but I did experience Nowruz: Persian New Year, celebrated on the spring equinox. My little sister is Iranian, and I was lucky to grow up surrounded by dinner tables full of Persian dishes.

Ghormeh sabzi was always my favorite: a comforting, cozy lamb and kidney bean stew that feels positively spring-y, thanks to an amount of fresh herbs you truly have to see to believe. Inspired by that dish, I wanted to create a centerpiece that could feed as many people as my sister and I now invite to our own Nowruz party: Stuffed & Roasted Leg of Lamb with a Mountain of Herbs.

To go alongside this lamb showstopper, we have Sabzi Polo with Tahdig—the most delicious herbed rice with a crispy crust everyone fights over—and A Super Fresh Yogurt Side. Because this Big Night falls on the more labor-intensive side, I like to bring in backup for dessert. The best baklava I've ever eaten is from Shatila—an iconic, decades-old

Middle Eastern bakery in Dearborn, Michigan, that, lucky for all of us, ships nationwide. Busting out this big gold box of flaky sweet-but-not-too-sweet pastry is the perfect note to end any meal on, especially this meal. Just drop it on the table for all to enjoy for the rest of the night (or until the box is empty).

Menu

For 8 to 12

SNAP PEA 'TINIS

RICOTTA TOASTS FOR EVERY MOOD

SABZI POLO WITH TAHDIG

STUFFED & ROASTED LEG OF LAMB WITH A MOUNTAIN OF HERBS

A SUPER FRESH YOGURT SIDE

BAKLAVA, FROM SHATILA

RICOTTA TOASTS FOR EVERY MOOD

Makes about 16 toasts

Yes, I am slathering ricotta on bread and calling it an appetizer. Groundbreaking? Of course not. But just like "Raspberry Beret" or "This Must Be the Place," this is one of my house's greatest hits. And I'm here to help you discover your own! If you (and your friends) love something, make it a signature. Especially when it's as easy—and as riff-able—as this.

1 baguette, cut into ½-inch-thick slices on a diagonal (more surface area for toppings!)

Extra-virgin olive oil

8 ounces whole-milk ricotta

Flaky sea salt

Toppings of your choice (see below)

Preheat the oven to 400°F.

Arrange the baguette slices on a baking sheet, then lightly brush the tops with olive oil. Toast for about 10 minutes, or until just golden brown.

Remove the toasts from the oven. Drizzle with a bit more oil, then generously dollop ricotta on top of each piece, schmearing it down slightly. Sprinkle with flaky salt, add whatever toppings you like, and serve.

• •

A Topping for Every Mood

- **ZESTY:** Grate lemon zest over each toast, then grind some black pepper on top and drizzle with honey.

- **SPICY:** Drizzle with homemade or store-bought hot honey and sprinkle with Aleppo pepper.

- **SWEET-TART:** Drizzle with pomegranate molasses.

- **FRESH:** Blanch 1 cup shelled English peas, then puree them in a food processor with olive oil and salt. Spread the puree directly on your toasts (before the ricotta), then drizzle with olive oil and sprinkle with cracked black pepper to finish.

- **BRIGHT:** Halve 1 pint cherry tomatoes, then toss with olive oil, salt, and black pepper to season. If you have time (or while the bread toasts), let the tomatoes sit for a few minutes, then drain any liquid. Scatter the tomatoes on top of the ricotta.

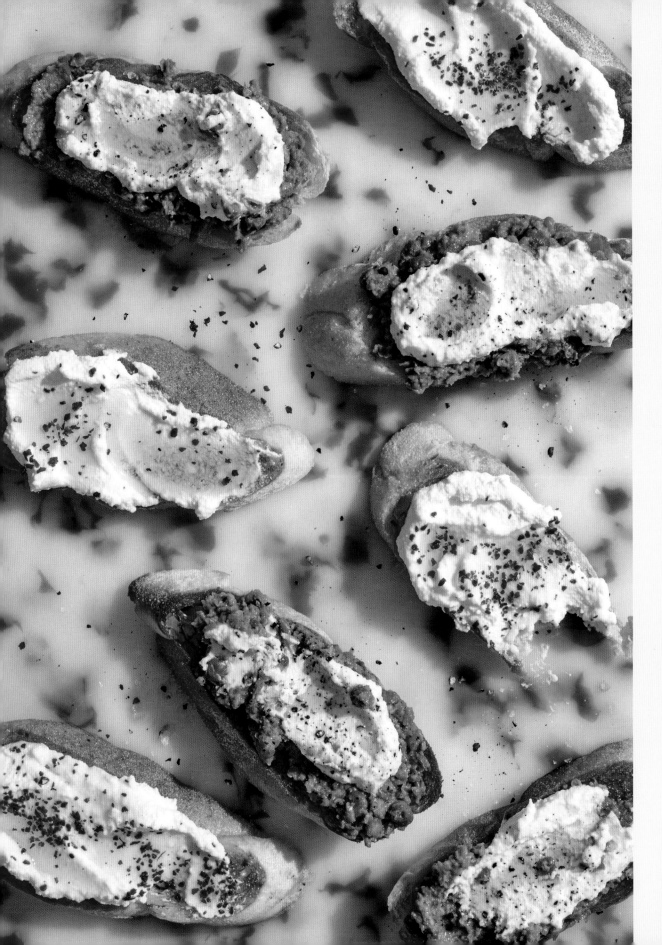

SABZI POLO WITH TAHDIG
with Roya Shariat

Serves 8 to 12

3 tablespoons fine salt

3 cups basmati rice, washed and rinsed thoroughly, soaked for at least 30 minutes, then drained

¼ teaspoon ground turmeric

1 teaspoon freshly ground black pepper

1 cup finely chopped fresh flat-leaf parsley leaves and tender stems

1 cup finely chopped fresh cilantro leaves and tender stems

1 cup finely chopped fresh dill

6 garlic cloves, finely chopped

4 tablespoons vegetable oil

2 limes, cut into wedges, for serving

NOTE: To achieve that covetable golden brown tahdig crust, a nonstick pot or skillet—or a very well-seasoned cast-iron pan—is essential here.

PAIR WITH: *Party Chicken with Feta & Fennel (page 27), Slow-Roasted Shawarma-Spiced Salmon (page 75), Not-Trendy, Actually Delicious Turkey & Gravy (page 207)*

Roya is a Big Night customer turned dear friend—someone I feel so lucky to have met through the shop. She's one of those people who seemingly does it all: She's an absolute powerhouse in her career at one of the world's largest beauty brands, she's a bright star of a human being who EVERYONE wants to be around, she makes killer playlists, she's a dog mom, and, oh, she's a cookbook author. *Maman and Me*, cowritten with her mother, Gita Sadeh, celebrates the evolution of Iranian food in America and is full of recipes perfect for hosting.

This versatile rice dish is a special treat typically eaten on Nowruz—the greens in the rice represent growth and renewal—and is most commonly paired with lamb or fish, but you could really serve it with anything (I love it with salmon, and I also love it tossed into a salad). Roya's family squeezes limes all over the rice, which isn't traditional, but makes the flavors truly shine.

Fill a large pot with a lid half full with water and bring to a boil over high heat. Once boiling, add the salt. Add the rice, turmeric, and pepper and stir to combine. Cover and cook the rice for 8 to 10 minutes, checking it at the 4- or 5-minute mark to see how quickly it's cooking. Taste one or two grains for doneness: The inside of the rice should be firm, while the outside of the grain should be cooked. With 1 minute of cooking left, stir in the parsley, cilantro, and dill.

Bring the pot to the sink and drain the rice into a colander. Rinse it thoroughly with cold water to stop the cooking process and let it drain completely. Transfer the rice to a medium bowl and stir in the garlic. Set aside.

In a small bowl, combine 3 tablespoons water with 2 tablespoons of the oil and set aside. In a medium nonstick pot or skillet, or a well-seasoned cast-iron pan, over medium-high heat, add the remaining 2 tablespoons oil and swirl to coat the bottom (you could use the same pot you parboiled the rice in, but a smaller pot will help you achieve a taller rice cake). Spread a thin layer of the parboiled rice on the bottom of the pot, gently patting it down to make a flat, even layer. Using a large spoon, gently layer the remaining parboiled rice on top, arranging it into a pyramid shape, so it doesn't stick to the sides of the pot.

With the handle of a wooden spoon, make four holes in the rice, going all the way to the bottom of the pan. Divide the reserved oil and water mixture among the holes. Wrap the pot lid with a kitchen towel and cover the pot tightly. Cook for 5 minutes to let the tahdig crisp up. Reduce the heat to medium and let the rice

steam for 35 minutes, or until it is fully cooked and you can hear the tahdig moving on the bottom of the pan when you shake it. If you start to smell something burning, you've overcooked it. The tahdig will turn a dark brown because of the herbs in the rice. Do not worry; it will still taste amazing!

Uncover the pot and place a large plate or platter over the top. In one smooth maneuver, quickly flip over the pot so the rice—including the tahdig—slips out onto the plate. If everything has worked out, you should have a gorgeous, crispy cake of rice. If not, don't worry! It will still be delicious. Use a spatula to grab any tahdig stuck to the bottom and scatter it over the top of the rice.

Serve with the lime wedges alongside for squeezing.

STUFFED & ROASTED LEG OF LAMB WITH A MOUNTAIN OF HERBS

Serves 8 to 12

2 tablespoons cumin seeds

2 tablespoons coriander seeds

2 tablespoons whole
   black peppercorns

2 tablespoons dried fenugreek
   leaves (or 1 tablespoon each
   coriander seeds and yellow
   mustard seeds)

4 ounces fresh parsley (about
   1 bunch)

4 ounces fresh cilantro (about
   1 bunch)

3 ounces fresh dill (1 to
   2 bunches)

2 ounces fresh chives
   (1 to 2 bunches)

3 ounces fresh mint leaves
   (from about ½ bunch)

6 garlic cloves, smashed
   and peeled

Kosher salt

10 tablespoons extra-virgin
   olive oil, plus more as needed

1 (4- to 5-pound) boneless leg of
   lamb (sirloin end), butterflied

3 or 4 medium leeks
   (1 to 1½ pounds)

1 lemon

Freshly ground black pepper

Flatbreads (page 74, or
   store-bought) pitas, or
   cooked rice, for serving

This recipe uses a boneless leg of lamb from the sirloin end, shank removed and butterflied. If that sentence just gave you anxiety, what I want is for you to call your butcher and ask for exactly that—they'll know what to do. Calling your butcher in advance ensures they have exactly what you need, and—if you're lucky—will have it prepped and waiting for you to pick it up. Thank you, butchers!

This recipe also calls for an herb called fenugreek (we're using the dried leaves, not the seeds), and while it may be a bit harder to find (you may need to order it online), I recommend the extra effort here. If you don't, this dish will absolutely still be delicious—but just not with quite the same ghormeh sabzi–inspired flavor. Either way, I hope you have as much fun making this lamb as I do. It's the most time-intensive main in the book, but it's also the kind of thing you will feel immensely proud to bring out and serve—and maybe even parade around the dining room first.

In a dry medium skillet, combine the cumin, coriander, and peppercorns and toast over medium heat, swirling the pan, until fragrant, 1 to 2 minutes. Remove the pan from the heat and let cool slightly.

In a spice grinder or a mortar combine the toasted spices and the fenugreek and grind (or crush with the pestle) into a coarse powder. Transfer the spice mix to a small bowl. *The spice mixture can be made up to 1 week in advance and stored in an airtight container at room temperature.*

Roughly chop half the parsley, cilantro, dill, chives, and mint (for a total of 3½ to 4 cups of herbs), reserving the remaining herbs whole (wrap them in dry paper towels and refrigerate).

In a food processor or blender, combine the garlic, ¾ teaspoon salt, and 6 tablespoons of the olive oil, then add a big handful of the chopped herbs. Pulse or blend on low speed until a paste begins to form, then add the remaining chopped herbs, 1 cup at a time, until a chunky paste forms (stir or scrape down the sides of the machine and add up to 1 tablespoon more oil to help it along, if needed). You should have about 1 heaping cup. *The herb paste can be made up to 1 day in advance and stored in an airtight container in the refrigerator.*

Add the herb paste to the bowl with the spices and stir to combine.

Place the lamb on a large cutting board or baking sheet. If purchased wrapped and rolled, cut off the twine and unroll the lamb. Pat both sides dry with paper towels. Remove any excess fat from the fleshy side (this can make lamb taste especially gamy). If the meat is thicker than 1 inch, cover the meat with plastic wrap and whack it with a rolling pin or meat tenderizer until it's ¾ to 1 inch thick. Season both sides well with salt (about ¾ teaspoon salt per pound of meat) and position the lamb fleshy-side up.

Smear about two-thirds of the spiced herb paste over the surface of the lamb. Working from one shorter, smaller end, tightly roll up the lamb from left to right. Use kitchen twine to truss the lamb at 1-inch intervals both lengthwise and crosswise. Smear the remaining spiced herb paste all over the outside of the lamb. Let sit at room temperature for 30 minutes, or transfer to a baking sheet or large plate and refrigerate, uncovered, for at least an hour or up to 12 hours.

Preheat the oven to 325°F.

Trim the roots and firm dark green tops from the leeks, then thinly slice the leeks and place in a large bowl of water to soak for a couple of minutes. Drain the water and repeat soaking and rinsing until the water runs clear. Strain the leeks and pat dry, then place them in a 6- to 8-quart Dutch oven along with the remaining ¼ cup oil and a big pinch of salt. Nestle the lamb into the leeks.

Cover and roast the lamb for 55 to 70 minutes, until fragrant and nearly cooked through (at least 120°F on an instant-read thermometer plunged into the thickest part; start checking it at 50 minutes, then at 15-minute intervals after that).

Uncover the lamb and increase the oven temperature to 500°F. Roast for 15 to 30 minutes more, until the lamb is deeply brown (130°F for medium-rare or up to 155°F for well-done, keeping in mind the internal temperature will rise at least another 10 degrees when the lamb comes out of the oven). Remove the lamb from the oven and let cool for 10 minutes. Transfer to a cutting board to rest for at least 30 minutes, reserving the leeks and pan juices.

Thickly slice the lemon, remove the seeds, then very finely chop the slices. Stir the lemon into the leek mixture and season with salt and pepper to taste.

When you're ready to serve, take the reserved herbs out of the fridge and roughly chop the leaves and tender stems. Reserve a handful and add the rest to the lemon-leek mixture, stirring to combine.

Snip off and discard the twine from the lamb. Slice the lamb thickly, adding some to each guest's plate and topping it with a few spoonfuls of the lemon-leek-herb mixture and a sprinkling of the reserved herbs. Bring the rest of the lamb to the table on a serving platter and serve with the remaining leek mixture and flatbread or rice alongside.

MAKE-AHEAD:

Spice mixture—Up to 1 week

Herb paste—up to 1 day

PAIR WITH: *Hidden Treasures Salad (page 198), Tahini-Miso Charred Greens (page 79)*

A SUPER FRESH YOGURT SIDE

Makes about 5½ cups; serves 8 to 12

Inspired by mast-o-khiar, a staple in many Persian meals, this yogurt is an endlessly riff-able, versatile side, that's also a dip, that's also a spread, or anything else you want it to be. Serve it with lamb, chicken, or salmon—it's great for a hit of freshness and acid to balance out the savoriness of pretty much any protein.

4 Persian cucumbers, diced

Kosher salt

4 cups (32 ounces) plain full-fat Greek yogurt or labneh

¼ cup extra-virgin olive oil

2 garlic cloves, peeled, plus more for serving

1 lemon, halved

½ cup thinly sliced scallions (from 4 to 6)

¾ cup packed chopped mixed fresh herb leaves, such as mint, dill, oregano, and tarragon

½ teaspoon dried mint (optional)

½ teaspoon dried dill (optional)

In a colander lined with a paper towel or kitchen towel, toss the cucumbers with a couple big pinches of salt.

In a large bowl, combine the yogurt, olive oil, and 1¼ teaspoons salt. Grate the garlic directly into the bowl. Squeeze the juice from half the lemon into the bowl. Add the scallions and chopped herbs, along with the dried mint and dill, if using. Stir to combine well.

Using the kitchen towel, scrunch the cucumbers to remove any excess water. Fold the cucumbers into the yogurt mixture. Taste and adjust the salt and lemon juice to your liking.

Cover and refrigerate for at least 30 minutes or up to 12 hours to let the flavors meld. Taste again just before serving and grate in additional garlic or add more lemon juice if you'd like.

MAKE-AHEAD: *Up to 12 hours*

PAIR WITH: *Party Chicken with Feta & Fennel (page 27), Slow-Roasted Shawarma-Spiced Salmon (page 75)*

SNAP PEA 'TINIS

Makes 1 'tini

2¾ ounces Snap Pea–Infused Gin (recipe follows)

¾ ounce dry vermouth (I like Dolin)

1 whole sugar snap pea, for garnish

NOTE: Place your glasses in the freezer before anyone arrives so they're chilled and ready for a round of 'tinis.

MAKE-AHEAD: *Infused gin— up to 1 week*

My favorite thing about spring is, undoubtedly, the arrival of sugar snap peas. I will eat a sugar snap pea any way, anytime, anywhere. Raw while walking down the street, charred on top of labneh for a quick springtime side, or dunked in literally any dip I can find, with abandon. One day, it dawned on me: *Why am I not eating my beloved snap peas in cocktail form?* My resident bartender and husband, Alex, promptly developed this drink, which combines two of my everlasting loves: snap peas and martinis. If you ask me about my soulmate, I might just say it's this cocktail (oh, and probably Alex, too).

Place your glass in the freezer to chill, if you haven't already. In a cocktail shaker filled halfway with ice, combine the infused gin and the vermouth. Cover the shaker and shake vigorously until very, very cold, about 20 seconds. Grab your glass out of the freezer. Pour the martini into the chilled glass and garnish with a snap pea.

Snap Pea–Infused Gin

Makes 10 ounces

8 ounces sugar snap peas, roughly chopped

10 ounces gin (I like Tanqueray)

Plop the chopped snap peas into a jar and pour in the gin. Seal, shake for about 10 seconds, and leave on the counter to infuse for 3 hours, or refrigerate for up to 24 hours. Strain through a fine-mesh sieve into a clean jar and seal. *Store the infused gin in the refrigerator for up to 1 week.*

SUNDAY IS FOR TACOS

IN THE RANKINGS OF BEST DINNER plans for a group, tacos will always be near the very tippity top for me. And for good reason. There are endless options, they're easily customizable to accommodate people's preferences and aversions, and, most important, they're delicious.

This particular Taco Night, however, is, in my humble opinion, a truly special one. Even better, it falls into the category of "seriously impressive food that requires far less work than your friends will think it did." Your pork can braise while you watch TV, do laundry, call your long-distance friend . . . in other words, it's an ideal Sunday project-cooking journey, with the most delicious taco spread as its finale.

And don't forget your tortilla plan. Though I definitely don't expect you to make your own tortillas, they're a crucial element of this Bigger Night! I love flour tortillas here, but if you're gluten-free, corn will be great, too. It's worth seeking out the very best tortillas you can for any taco night—and because we're not all blessed with the best grocery store tortillas (looking at you, Texans, and your H-E-B), note that Caramelo makes incredible ones—and they ship nationwide.

This spread is a perfect way to end any week, and—bonus—you'll be giving yourself the gift of al pastor-ish leftovers for your Monday.

Menu

For 8 to 12

IMPROVED MEZCAL PALOMAS

CHIPS & TOMATILLO-AVOCADO SALSA

BRAISED AL PASTOR-ISH TACOS

TRE LATTI CAKE

IMPROVED MEZCAL PALOMAS

Makes 2 cocktails

3 to 4 jalapeño slices, seeded

5 ounces fresh grapefruit juice (from 1 large grapefruit)

2 ounces fresh lime juice (from about 2 limes)

4 ounces mezcal (I like Del Maguey Vida)

1 ounce agave nectar

1½ ounces triple sec (I like Giffard Curaçao, triple sec, or Cointreau)

2 to 4 ounces sparkling water

2 grapefruit slices, for garnish

This cocktail takes inspiration from the Paloma, which is the most famous example of why and how grapefruit and tequila are best friends forever. To be clear, the Paloma doesn't need improvement. This recipe just showcases how I like mine: mezcal swapped for tequila; the addition of jalapeño; and instead of sparkling grapefruit soda, a combination of fresh juice and sparkling water. I developed this concoction while on vacation, and, thus, it always brings me back to the feeling of a hot, lazy afternoon spent by a pool, overlooking the ocean. Give your friends a little bit of that energy by serving them this drink.

In a cocktail shaker, mash the jalapeño slices with a muddler or wooden spoon for a few seconds, breaking them down just a bit to bring out their flavor.

Add the grapefruit juice, lime juice, mezcal, agave, and triple sec. Add ice, seal the shaker, and shake vigorously for 30 seconds. Strain (I like to double-strain this one through a fine-mesh sieve) into two highball glasses filled with ice, dividing evenly. Top with sparkling water and garnish each glass with a grapefruit slice.

TOMATILLO-AVOCADO SALSA

Serves 8 to 12 (makes about 2 cups)

This salsa is extremely versatile—it's up for pretty much any adventure you want to have with it. Leave it chunky, or try a smoother, blended version if that's what you're into. Try it on tacos, salads of all kinds, fish, chicken, or even eggs. It has a real depth of flavor, without stealing the show from anything else you pair it with. You can even skip the avocado if you can't find ripe ones. (Just try it with the avocado at least once, okay? For me.)

2 tablespoons neutral oil

6 to 8 tomatillos, husks removed, rinsed well, and chopped

1 small red or yellow onion, chopped

1 (4-ounce) can chopped green chiles

¼ teaspoon sugar

Kosher salt and freshly ground black pepper

2 garlic cloves, grated

1 large jalapeño, chopped (optional)

2 ripe avocados

½ bunch cilantro

1 lime

Tortilla chips, for serving

Heat the oil in a large skillet over medium heat. When the oil is shimmering, add the tomatillos, onion, chiles, and sugar. Season with a big pinch of salt and lots of pepper. Cook, stirring occasionally, until the tomatillos are softened, about 10 minutes. Add the garlic and cook, stirring frequently, until fragrant, about 1 minute more. Transfer to a medium bowl and let cool to room temperature (about 20 minutes; if you want to speed this up, you can place the bowl in a larger bowl filled with ice and water for 10 minutes).

Stir in the jalapeño (if using). Let sit for at least 10 minutes or up to 4 hours.

Peel and roughly chop the avocados, then add them to the bowl. Finely chop the cilantro, then stir it into the bowl. Season with a big squeeze of lime juice and more salt and pepper, until the flavors pop. Enjoy with chips and/or on tacos.

NOTE: This recipe is for a chunky salsa. If you're in a smooth mood, blend the tomatillo mixture and cilantro before adding the chopped avocado.

MAKE-AHEAD: *Up to 4 hours*

BRAISED AL PASTOR-ISH TACOS

Serves 8 to 12

The three main ingredients for this al pastor-inspired dish are pork, time (4 to 6 hours, all in), and your trusty oven. Al pastor is traditionally cooked on a rotating vertical spit called a trompo, which is absolutely on my Dream Cooking Tool Wish List, but it's an item most of us don't have lying around the house. Instead, this pork spends the day simmering in a luxurious bath while your time is freed up to do the same—or to make Tomatillo-Avocado Salsa (page 61), or to spend your Sunday however else you want.

FOR THE BRAISED PORK

6 garlic cloves, peeled

1 (7-ounce) can chipotle peppers in adobo sauce

2 tablespoons chile powder (guajillo or ancho, preferably)

2 tablespoons smoked Spanish paprika

1 tablespoon ground cumin

6 tablespoons neutral oil

1 (3½- to 4-pound) boneless pork shoulder, strings and/or skin removed

Kosher salt

1 large white onion, quartered

2 cups cubed pineapple

2 tablespoons tomato paste

½ cup apple cider vinegar

FOR SERVING

Small corn or flour tortillas, warmed

1 cup cubed pineapple

Diced white onion

Chopped fresh cilantro

Sliced radishes

Lime wedges

NOTE: Just before serving, if you're so inclined, feel free to put your entire Dutch oven under your broiler for 1 to 2 minutes. You'll get some crispier bits of pork on top of the rest of the saucy, juicy braise.

MAKE THE PORK: Smash 3 of the garlic cloves and set aside. Grate the remaining 3 garlic cloves into a small bowl. Add ¼ cup of the adobo sauce from the can of chipotle peppers, the chile powder, paprika, cumin, and 3 tablespoons of the oil. Stir together until a paste forms.

Place the pork shoulder on a baking sheet or large plate. Season all over with salt (1 teaspoon per pound of meat is a good rule of thumb), then smear the spice paste all over the surface of the meat. Let sit at room temperature, uncovered, for at least 30 minutes and up to 2 hours.

Position a rack in the center of the oven and preheat to 350°F.

Heat the remaining 3 tablespoons oil in a 6- to 8-quart Dutch oven over high heat. When the oil is shimmering, add the onion and 1 cup of the chopped pineapple to the pot and cook, stirring occasionally, until it's browned on all sides, 7 to 9 minutes. Reduce the heat to medium and scoot the onion and pineapple to one side of the pot. Add the remaining garlic cloves, the tomato paste, and 3 chipotle peppers. Stir to combine, then let sizzle until the mixture darkens a bit and becomes sticky, about 45 seconds. Stir in the vinegar, season with a couple big pinches of salt, and remove from the heat.

Nestle the pork into the center of the pot so that one of the widest sides is lying flat (fat-side up, if possible). Add ½ cup water, cover, and transfer to the oven. Braise for 1 hour, then remove from the oven and spoon some of the liquid over the pork. Cover and return to the oven. Continue to braise the pork, basting and adding more water as needed every hour or so, until the meat is tender and shreds when pulled with a fork, another 2 to 3 hours.

MAKE-AHEAD: *Pork—up to 2 days*

PAIR WITH: *The Corn Salad I Think About All Year (page 113)*

Increase the oven temperature to 425°F. Top the pork with 1 cup chopped pineapple and return to the oven, uncovered. Roast for 20 to 25 minutes, until some of the liquid has reduced, the added pineapple is slightly tender, and the pork is super tender. Remove from the oven and use two forks to shred the pork, then stir to incorporate with the pan juices. *The pork can be stored in an airtight container in the refrigerator for up to 2 days. To* *reheat, add a splash of water to the pot, stir, cover, and cook over medium-low heat on the stove or in a 350°F oven until warmed through.*

Serve the pork in the tortillas with the chopped pineapple, raw onion, cilantro, and radishes for topping, and lime wedges alongside for squeezing. If you happen to have any leftovers, the pork is also amazing in a rice bowl or on tostadas.

TRE LATTI CAKE

Serves 8 to 12

FOR THE CAKE

Nonstick cooking spray

2¾ cups all-purpose flour

2 teaspoons baking powder

2 teaspoons kosher salt

5 large eggs

1 (14-ounce) can sweetened
    condensed milk

½ cup sugar

½ cup neutral oil

FOR ASSEMBLY

3 cups brewed coffee or
    espresso, at
    room temperature

1 (14-ounce) can sweetened
    condensed milk

¾ cup dark rum or medium
    amaro, such as Montenegro
    or Meletti (optional; replace
    with additional coffee if
    omitting—see headnote)

2 cups cold heavy cream

2 cups mascarpone

¼ cup sugar

2 teaspoons pure
    vanilla extract

1 teaspoon kosher salt

1 tablespoon unsweetened
    cocoa powder (preferably
    Dutch-process) and/or
    1 teaspoon ground cinnamon

After debating for months and months whether to purchase myself a fancy espresso machine, I finally bit the bullet and ordered one. When it arrived, I did what any reasonable person would do: I threw a party. My friend Skyler made a tres leches cake for the occasion, which was when we had the (frankly, brilliant—and also brilliantly caffeinated) idea to drown the cake in espresso. After that, I knew everyone deserved to experience the blissful union of tiramisu and tres leches. The result is this cake, which is an ideal make-ahead dessert, as it should really sit in the refrigerator for a full 24 hours before you serve it.

Regarding the booze—using rum will give you a slightly sweeter cake, while amaro will lend a slightly more bitter flavor (I've noted my favorite options in the ingredient list). If you'd prefer to go booze-free, just skip it and add more coffee in its place.

MAKE THE CAKE: Position a rack in the center of the oven and preheat to 350°F. Coat a 9×13-inch cake pan (see Note) with nonstick spray, then line it with parchment paper.

In a medium bowl, whisk together the flour, baking powder, and salt.

Crack the eggs into a large bowl. Add the sweetened condensed milk and the sugar and, using a hand mixer, beat on medium speed until combined, about 1 minute. Reduce the speed to low and add half the flour mixture, mixing until just combined. Add the oil, then the remaining flour mixture, and mix until just combined. Scrape the batter into the prepared pan. Smooth out the top with a spatula.

Bake, rotating the pan halfway through, until golden and a tester inserted into the center comes out clean, 25 to 30 minutes. Let the cake cool for 15 minutes in the pan, then turn it out onto a wire rack to cool completely, about 2 hours. *The cooled cake can be tightly wrapped and stored at room temperature for up to 3 days or in the refrigerator for up to 5 days.*

TO ASSEMBLE: Poke the cake all over with a fork, making sure to get all the way to the bottom. Split the cake in half horizontally (like a layer cake). Place half the cake, cut-side up, in a clean 9×13-inch cake pan or another 3- or 4-quart baking dish (if the dish isn't rectangular, just break pieces of cake to fit in an even layer). You may need to trim the edges of your cake so that it fits flat in the bottom of your dish.

In a small bowl, stir together the coffee, condensed milk, and rum (if using). Pour about half the mixture over the cake, or as much as you need to cover it completely. Let soak for at least 10 minutes. If you notice parts that seem a little dry, poke more holes in the cake at this point and add a little more liquid.

Meanwhile, in a large bowl, combine the cream, mascarpone, sugar, vanilla, and salt. Using a hand mixer, beat on medium speed until the cream holds medium peaks, 1 to 2 minutes.

Spread half the whipped cream mixture over the top of the soaked cake. Place the second half of the cake, cut-side up, on top of the whipped cream, then pour over more coffee mixture to completely soak the cake. (You may have more liquid than you need—do not pour so much that you create a moat around the cake. You could use what's left to make a side cake with your cut-off cake edges!) Top with the remaining whipped cream.

Chill the assembled cake in the refrigerator for at least 6 hours, or ideally 24 hours, before serving. (It will keep, covered, in the fridge for up to 4 days.) In a small bowl, stir together the cocoa powder and cinnamon, then use a fine-mesh sieve to dust the chilled cake with the mixture.

NOTE: If you don't have a 9 × 13-inch cake pan, bake the cake on a 13 × 18-inch rimmed baking sheet—*not* in a 9 × 13-inch glass or ceramic baking dish—and start checking it after 18 minutes in the oven (it will bake faster). However! A baking dish *will* be best to assemble the final, stacked cake in, so have that handy, too. If you're using a baking sheet, cut the cake crosswise into two 9 × 13-inch rectangles when you're ready to assemble.

MAKE-AHEAD:

Baked sponge cake—up to 5 days

Assembled cake—up to 4 days

stop worrying about everything matching · · · · · · · · · · · · · ·

Here are things nobody will ever say:
Why don't all your plates match?
Wait, your place mats aren't all the same color?
How come you only have four of this kind of fork but eight of the other?

Here's what they will say:
Wow, this table looks amazing.
These colors!!!
I've been looking for place mats/forks/plates like this. Where did you find them?!
No. Really. *Where?*

In other words: No one cares if your place settings match. And if they do . . . that's their thing, and it certainly doesn't need to be yours. What people do notice is the effort you put into setting the table. How you made it feel warm and inviting, like you wanted people to sit down, get comfortable, and enjoy. So stop worrying about the matching thing! Find pieces you love that bring you joy—full stop. Flip through the tables shown in this book for a little inspiration, and I can guarantee you'll see new ways of putting together the "mismatched" things you already have.

→ Don't know where to even start? First, consider the logistics:

WHERE?

Start from the ground up—where are we eating?

WHO?

Who's coming over? Will we need extra chairs? How much space will we actually have on the table?

WHEN?

When did you tell people to show up? The table should be set before anyone arrives. Add it to your Hosting Timeline.

WHAT?

What are we serving, and is it buffet-style or on the table? What serving pieces do we need?

HOW?

How are we eating? Do we need a plate for salads? Could everything go on one plate? Or a blate (bowl-plate)?

→ Then, consider some aesthetics:

VIBE

Are we doing a snacking table, where individual place settings aren't required? Are we looking for glamour, or is this a laid-back weeknight hang?

MOOD

Peak-summer fresh? Wintry cozy? Springy and bright? It's your Big Night—it should absolutely reflect your mood.

... and just set the table

→ Now, let's build this table, layer by layer:

BASE LAYER

In order of formality:

1. Naked table (no linens)
2. Just place mats
3. Tablecloth and place mats

ESSENTIALS LAYER

Your plates, bowls, or whatever people will be eating off of. Once again: Use what you have and don't worry if it doesn't all match.

GLASSWARE LAYER

Everyone gets a water glass. And a wine glass, if you're serving, for whoever might want some. You don't need to set cocktail glasses—people will be holding those from the beginning of the evening. For me, the glassware layer is where the fun really starts—and the more mix of color, style, and shape, the better.

FLATWARE LAYER

The order you lay out your forks, spoons, and knives simply does not matter. All that matters is that everyone has them.

LIGHTING LAYER

As Cher Horowitz famously explained in *Clueless*, we need to "design a lighting concept." Try to avoid overhead lighting, opting instead for lamps and/or lighting on the table instead. Candles are always welcome if you've got room—tapers bring the drama, while votives are more low-key.

SEASONING LAYER

Don't forget your salt, pepper, butter, chili crisp, olive oil, or whatever else you want within arm's reach.

PERSONALIZED LAYER

Place cards can be a special touch for a larger group if you want to have fun with a seating arrangement and/or introduce your friends to one another.

X-FACTOR LAYER

Entirely unnecessary, but if you're feeling it . . .

- Whole citrus or fruit—sparingly, please; no one wants to elbow a lemon off the table
- Leaves, such as eucalyptus
- Crayons and coloring paper—discover your inner child
- Flowers in bud vases or a bigger (but short) arrangement, so no one has to crane their neck to talk to people

YOUR BEST FRIEND'S BIRTHDAY

YOU NEVER FORGET YOUR FIRST DINNER PARTY—that first time you personally took it upon yourself to plan a menu, set the table, open your door, and cook for people you really, really like.

I can tell you whom I invited (a mix of work friends and college friends) and what I was wearing (a vintage, off-the-shoulder dress that I thought perfectly said "oh, what, this old thing?"). I can tell you where we sat (at my pink dining table, which I still use today, purchased for $15 at my first job in the corporate offices of J.Crew when there was a store props sale). And, of course, I can tell you exactly what we ate (chicken shawarma made from a *New York Times* recipe that had so many five-star reviews, I was confident I could be one of them).

But what I remember most viscerally about that night is the feeling of bringing together a group of people who might not otherwise be together, and spending the evening eating, drinking, and talking without a care in the world as to when we might have to get up from the table, pay the check, or figure out where we were going next. This feeling is one of the greatest gifts you can give—especially to a best friend. Especially for their birthday.

For my best friend's birthday dinner, she gets glamour—in the form of a big, beautiful piece of fish (no one needs to know just how easy it is to cook it). She gets Negronis—not just one way, but two. And she gets a sundae, with her favorite ice cream and homemade fudge. Most of this dinner's key elements come together in advance, so once it's time to party, you can spend less time cooking and more time toasting your BFF.

Menu

For 6 to 8

**MARINATED OLIVES
FOR ANYTIME**

**NEGRONIS, 1 OR 2 WAYS
(SEE PAGES 29 & 31)**

FAST YOGURT FLATBREADS

**SLOW-ROASTED SHAWARMA-
SPICED SALMON**

**TAHINI-MISO
CHARRED GREENS**

**TAHINI HOT
FUDGE SUNDAES**

FAST YOGURT FLATBREADS

Makes 12 small or 6 large flatbreads

3 cups all-purpose flour

1 tablespoon baking powder

1 tablespoon kosher salt

1 cup plain full-fat
   Greek yogurt

Extra-virgin olive oil

NOTE: These flatbreads are the perfect blank canvas for sauces, vegetables, meats, and fish—but if you want to add a little something extra, combine 1 tablespoon of za'atar or ground sumac with 3 tablespoons of olive oil, then brush the mixture onto the warm flatbreads as soon as they're out of the skillet.

MAKE-AHEAD: *Up to 24 hours*

PAIR WITH: *Party Chicken with Feta & Fennel (page 27), Stuffed & Roasted Leg of Lamb (page 49), Lighthouse's Hummus (page 154), Creamy Tomato Soup (page 224)*

People are always, *always* impressed by fresh bread. Does that mean you need to make your own bread every time you have people over? Absolutely not. But if you're putting a menu together and you think to yourself, *I could really use a carby vehicle here,* remember these flatbreads. They come together quickly and easily and, unlike a loaf of sourdough, they're also relatively foolproof.

In a medium bowl, whisk together the flour, baking powder, and salt. Stir in the yogurt and ¼ cup water until the mixture forms a shaggy dough—if there are dry pockets of flour, add an additional 1 to 2 tablespoons water until it comes together. Use your hands to knead the dough in the bowl until well combined and almost smooth, about 5 minutes—it should be tacky but not sticky, like Play-Doh. Cover the bowl with a clean kitchen towel and let the dough rest for 20 minutes.

Turn out the dough onto a clean work surface and divide into 6 or 12 equal pieces (the dough should be significantly softer now). Cover with the kitchen towel. On a lightly floured surface, working with one ball at a time, roll out each ball of dough, flipping occasionally, into a roughly 7-inch (for larger flatbreads) or 5-inch (for smaller flatbreads) round, placing them on a baking sheet and stacking them as needed.

Heat a large cast-iron skillet over medium-high heat until it smokes, then oil the pan. Place a dough round in the skillet and cook until bubbles start to form on top, 1 to 2 minutes. Flip and cook another 1 to 2 minutes on the other side until golden brown all over. Transfer the finished flatbread to a basket or plate and cover with a kitchen towel or cloth to keep them warm while you make the rest. Repeat with the remaining dough rounds, adding a little oil to the pan as needed throughout the process.

These flatbreads are best when they're served immediately, but if you're serving them later, let the flatbreads cool, then wrap them tightly in foil and store at room temperature for up to 24 hours.

SLOW-ROASTED SHAWARMA-SPICED SALMON

Serves 8

1 (3-pound) center-cut salmon fillet, skin removed

Zest and juice of 1 lemon

Kosher salt and freshly ground black pepper

2 to 4 tablespoons Shawarma Spice Mix (recipe follows)

¼ cup extra-virgin olive oil

1 medium red onion, thinly sliced

FOR SERVING

Fast Yogurt Flatbreads (page 74), store-bought flatbreads or pitas, or rice

Sliced cucumber and/or radish, salted and drained

Yogurt and/or tahini (optional)

PAIR WITH: *A Super Fresh Yogurt Side (page 51), Sabzi Polo with Tahdig (page 46), Hidden Treasures Salad (page 198)*

Slow-roasted salmon is one of my dinner party go-tos that's guaranteed to impress. Conveniently, it's also exceptionally easy to pull off (but no one else has to know that). Once you've made the shawarma spice mix and procured your salmon—it should go without saying that the better the quality of fish, the more delicious this will taste—you're basically there. Serve it with a big spoon, which is the easiest way to get it out of the pan—and also helps keep this dish from feeling fussy.

Position a rack in the center of the oven and preheat to 325°F. Line a 2- to 3-quart baking dish or a rimmed baking sheet with parchment paper.

Place the salmon in the prepared baking dish. Drizzle 1 to 2 tablespoons of the lemon juice over the salmon to coat; reserve the remainder. Season the salmon liberally with salt and a few grinds of pepper, then cover completely with the shawarma spice mix. Drizzle the olive oil over the top.

Your cook time will vary based on the thickness of your fillet. If you're going for medium-rare (it flakes with a fork at the thinner parts but is still a bit opaque at the thickest parts), start checking the salmon after 15 to 20 minutes. If you're going for medium or medium-well (flaky all the way through), start checking it after 20 to 25 minutes. Immediately after removing from the oven, tilt the baking dish slightly and use a spoon to scoop up the rendered fat and spoon it over the salmon for 30 seconds. Let cool for 5 minutes, then scoop up large pieces of the fillet with a fish spatula or spoon (it will break naturally) and place on a serving platter. Pour the spiced oil from the pan over the salmon.

While the salmon roasts, in a medium bowl, combine the lemon zest and remaining lemon juice. Add the onion and a big pinch of salt. Let sit until you're ready to serve, at least 15 minutes.

Just before serving, spoon the lemony onions over the top of the salmon. Serve with flatbread or rice, as well as the cucumbers, radishes, and yogurt or tahini, if desired.

→

Shawarma Spice Mix

Makes about ½ cup

2 tablespoons
    ground coriander

2 tablespoons ground cumin

2 tablespoons sweet or hot
    smoked paprika

1 tablespoon garlic powder

2 teaspoons ground sumac

1 teaspoon ground ginger

½ teaspoon ground cinnamon

½ teaspoon ground cloves
    or allspice

¼ teaspoon cayenne pepper,
    or ½ teaspoon red
    pepper flakes

In a jar, combine the coriander, cumin, paprika, garlic powder, sumac, ginger, cinnamon, cloves, and cayenne. Seal the jar and shake to mix well. *Store in a cool, dark place with your other spices for up to 6 months (but I bet it won't last that long).* Now you're that much closer to your next Shawarma-Spiced Salmon Night.

MAKE-AHEAD: *Up to 6 months*

TAHINI-MISO CHARRED GREENS

Serves 6 to 8

Neutral oil

3 large bunches lacinato kale or another dark leafy green, stems removed and leaves roughly torn

Kosher salt and freshly ground black pepper

1 lemon or lime, halved

4 garlic cloves, peeled

Tahini-Miso Special Sauce (page 37)

Mild chile flakes, such as Aleppo pepper or gochugaru, for serving

Crunchy, Roasty Glitter (page 34), for serving (optional)

MAKE-AHEAD: *Up to 2 days*

PAIR WITH: *Party Chicken with Feta & Fennel (page 27), beef from Bulgogi-ish Lettuce Wraps (page 169), Crispiest Chicken Milanese (page 150)*

How to make a side of roasted or sautéed vegetables feel more exciting than . . . a side of vegetables? Put a generous swipe of something underneath them. Ricotta or labneh seasoned with olive oil, salt, and pepper, for example, is excellent underneath kale, broccolini or broccoli rabe, carrots, and beets. Tahini, cut with a little lemon, is another no-fail option. Even better, make a batch of Tahini-Miso Special Sauce, spoon as much as you'd like onto your platter, and top with leafy greens after quickly charring them in a skillet.

Heat 2 tablespoons oil in a 12-inch cast-iron skillet over medium-high heat. When the oil is shimmering, add a large handful of the kale to the pan, season with salt and pepper, and quickly toss to coat. Cook, undisturbed, until the kale begins to char on the one side, 3 to 5 minutes. Toss for 30 seconds, until mostly wilted, then transfer to a large bowl. Repeat with the remaining kale, adding more oil in between batches as needed. Juice half the lemon into the warm greens. Grate the garlic into the bowl and toss to combine. Taste and add more salt, pepper, and lemon juice as needed. *The greens can be stored in an airtight container in the refrigerator for up to 2 days.*

Smear a generous amount of tahini-miso sauce over the bottom of one or two shallow serving bowls, then add the greens. Top with chile flakes and, if desired, crunchy, roasty glitter. Cut the remaining lemon half into wedges and serve alongside for squeezing.

MARINATED OLIVES FOR ANYTIME

Serves 6 to 12

1 tablespoon whole
   black peppercorns

1 tablespoon fragrant seeds,
   such as fennel, cumin,
   or coriander

5 garlic cloves, smashed
   and peeled

3 long strips of lemon or
   orange peel

3 cups unpitted green olives,
   such as Castelvetrano,
   Cerignola, and/or picholine

⅓ cup dry-cured black olives
   (optional)

3 sprigs woody herbs, such as
   thyme, oregano, rosemary,
   or sage

1 cup extra-virgin olive oil, plus
   more for serving

6 ounces semi-firm cheese,
   such as Gouda or
   Manchego, or fresh cheese,
   such as feta or goat,
   crumbled (optional)

2 tablespoons fresh lemon
   juice, or red or white wine
   vinegar, for serving

Flaky sea salt

In my house, olives are welcomed, loved, and appreciated in any and every appetizer spread. They're the Little Black Dress of snacks—they work for any occasion, with any other dishes. All you really need to take a tub of olives from grocery store to glam is a great serving vessel, a drizzle of good olive oil, a few cracks of black pepper, and lemon zest. But if you have a bit of extra time or foresight, you should absolutely marinate them. This recipe is flexible—if you're missing an ingredient, don't sweat it. You can even skip the cheese altogether, although that's an easy way to fill out this dish and make it a heartier appetizer, especially when you add crusty bread.

In a spice grinder or with a mortar and pestle, gently crush the peppercorns and fragrant seeds.

In a medium saucepan, combine the crushed spices, garlic, citrus peel, olives, woody herbs, and olive oil and cook over medium-low heat, swirling the pan occasionally, until the garlic is golden and the mixture is warm and fragrant, 10 to 15 minutes. Remove the pan from the heat and transfer the mixture to a shallow serving bowl. Let cool slightly, about 15 minutes. *The mixture can be stored, covered, in the refrigerator for up to 1 week; remove the garlic cloves before storing. Reheat over low heat until warmed through, 5 to 10 minutes.*

If using cheese, place it in the serving bowl and gently toss to coat. Pour the citrus juice over the top and finish with a glug of good olive oil and a pinch of flaky salt before serving.

MAKE-AHEAD: *Up to 1 week*

PAIR WITH: *Literally any other appetizer, salad, or main dish in this book*

TAHINI HOT FUDGE SUNDAES

Serves 8 or more

FOR THE TAHINI FUDGE

½ cup (1 stick) unsalted butter

1½ cups agave nectar

1 cup heavy cream

1 tablespoon kosher salt

½ cup tahini

4 ounces bittersweet
chocolate, roughly chopped

FOR THE SUNDAES

Vanilla, chocolate, and/or
coffee ice cream, softened

1 (8-ounce) package halva
(optional)

Whipped cream

Shelled roasted, salted
pistachios, chopped
(optional)

Maraschino, amaro-soaked,
or pitted fresh cherries
(optional)

NOTE: If you have wide
coupes or any other glasses
that can accommodate ice
cream, use them. Seeing
the layers of hot fudge, ice
cream, and whipped cream
as you eat is an extra treat.

MAKE-AHEAD: *Tahini
fudge—up to 3 weeks*

If you have ice cream in your freezer, you already have a dinner party dessert. But when you want to go big(ger), set up a sundae bar. Make the tahini hot fudge in advance—then, when you're ready to serve, pull it out of the fridge and warm it in a saucepan over low heat. Grab the ice cream out of the freezer. While the fudge warms and the ice cream softens, make whipped cream. Gather some cherries and/or nuts and/or halva, and, suddenly, you have a dessert that everyone will be thinking about tomorrow.

MAKE THE FUDGE: Melt the butter in a large saucepan over medium heat. Whisk in the agave, cream, and salt. Whisking occasionally, bring to a rolling boil, 3 to 5 minutes. Continue to boil, whisking often and adjusting the heat as needed to keep the pot from boiling over, until the mixture has darkened slightly and thickened enough to coat the back of a spoon, 10 to 15 minutes.

Reduce the heat to low, then whisk in the tahini and chopped chocolate until smooth and incorporated.

Remove the pan from the heat and pour over ice cream immediately, or cover to keep warm (to reheat, simply place over low heat until you've reached peak pourability). *The fudge can be made up to 3 weeks in advance and stored in an airtight container in the refrigerator. Reheat in the microwave in 30-second intervals, or in a pot over low heat.*

MAKE THE SUNDAES: Scoop a bit of the fudge into bowls or cups. Add a couple of scoops of ice cream to each (only *you* know how much is "enough"). Drizzle with more fudge, crumble the halva over the top, dollop with whipped cream, sprinkle with pistachios, and, of course, top with a cherry.

BIGGER NIGHTS

vacation big nights   season to (your) taste
88   92

DIY BLT NIGHT

Serves 8

3 pounds bacon

6 to 8 large heirloom or
   beefsteak tomatoes

Kosher salt and freshly ground
   black pepper

1 lemon, sliced into wedges

1 to 2 heads lettuce (I like
   romaine or butter),
   leaves separated

1 loaf of bread, such as
   sourdough, rye, or white,
   sliced, or Fluffy Sheet Pan
   Focaccia (page 220), sliced
   in half through the middle

Sliced avocado (optional)

Mayonnaise or A Quick Aioli
   (page 99)

Chili crisp (optional)

NOTE: Don't skip the tomato-
draining step. After seasoning
with salt and pepper on both
sides, let them sit for at least
several minutes. The longer
they drain, the more their
peak-summer flavor intensifies.

PAIR WITH: *The Corn Salad I
Think About All Year (page 113),
Herby Double Summer Bean
Salad (page 110), A Chic! Potato
Salad (page 117), Crunchy, Creamy
Buttermilk Slaw (page 103)*

I came to tomatoes late in life. It wasn't until I tasted a Brooklyn-
rooftop-grown cherry tomato in my early twenties that I finally
understood what I had been missing out on. I've been making up
for lost time ever since. During this BLT-intensive period of my life, I
have come to know four rules for the best three-letter-sandwiches:

1. The tomatoes must be as good (and ripe) as you can get 'em.

2. The vegetables must be properly seasoned. Tomatoes: salt
   and pepper, both sides. Lettuce: olive oil, salt, and lemon juice.

3. There must be mayo—and lots of it. Slather the insides of your
   toast with it as though you are icing a cake.

4. The order of assembly matters. Tomatoes should go
   straight onto the mayo-covered bread. That's where the
   magic happens.

Preheat the oven to 425°F. Line two baking sheets with foil. Line a
large platter with paper towels.

Divide the bacon between the prepared baking sheets. Bake,
rotating the pans halfway through, until the bacon is brown and
crisp (please, no flabby bits!)—begin checking it around 10 minutes,
then every 3 to 5 minutes after that. Transfer the bacon to the
prepared platter to drain.

While the bacon is cooking, slice the tomatoes. Place on a paper
towel–lined plate or baking sheet, season all over with salt and
pepper on both sides, and set aside to drain. Peel and slice the
avocado (if using) and season with salt and a squeeze of lemon.

When you're ready to serve, transfer the bacon and tomatoes
to serving platters. In a medium bowl, season the lettuce with
salt and lemon juice, then set on a platter. Toast the bread. (Try
dipping the bread into the rendered bacon fat from the baking
sheet and toasting it in a skillet on both sides.) Set out the toast,
avocado, and remaining lemon wedges. Set out the mayonnaise,
swirling in chili crisp if you'd like it spicy, or make a quick aioli, if you
like it garlicky. For extra credit, serve it all three ways so everyone
can find their personal mayo bliss.

In case anyone needs a reminder on the correct order of
assembly: toast, mayo on both sides, tomato, bacon, lettuce. If
using avocado, add that to the other side of the toast. Close, slice
in half with a serrated knife, then devour.

vacation big nights

YEARS AGO, a group of friends and I made the time-honored pilgrimage from New York City to upstate New York to attend our wonderful friend Jeana's wedding. We had rented a house and would be driving up after work, arriving pretty late. Our dinner plan was to keep it easy: pasta with store-bought pesto, plus some grilled vegetables. (Backyard grilling is especially exciting when you live in a city where most people have to go to a park to grill.)

Two things we had not factored into our plan: a torrential downpour, which made for unsafe driving conditions that added a fun little layer of stress and exhaustion for everyone involved, and, once we finally got to the rental house, no gas for the stove—and no way to turn the gas on. Luckily, my friends are brilliant, and we pivoted. We put our pot of water right on the grill, and once it was boiling, cooked the pasta as though it was on the stove. And you know what? The dinner worked perfectly fine.

The point of this story is not just that you can, in fact, boil water on a grill. The point is that meal planning is essential for Big Nights on vacation. It's also essential for you to expect that plan to not go exactly according to plan. Humans are imperfect, and so are vacation rental homes.

Like the Hosting Timeline (page 32), planning your Vacation Big Nights in advance of vacation means you get to fully enjoy yourself *on* that vacation, instead of wondering what you should eat. Sure, scheduling vacation meals in advance doesn't sound very . . . vacation-y. But who wants to be sitting on a beach, or hiking, or reading a book, while there's a little voice in the back of your head wondering, *What's for dinner? Who's going to make it? Do we need to go to the grocery store?*

Meal planning also doesn't need to mean every single food moment is accounted for. Leave half the nights of your trip open to spontaneity, if that's what feels right! But it is my firm belief that some of the best vacation memories happen around a dinner table, and planning out a few of those meals can't hurt. Here are some suggestions:

A BIG WEEK UPSTATE WITH SIX FRIENDS ·

Saturday—Arrive late afternoon, need something easy and low-effort . . .
- BLT Night
- Corn & Strawberry Pop-By Cake (page 104; make in advance and bring with you)

Sunday—Grilling & Chilling
- Grilled Chicken Sandwiches with Melted Swiss & Slaw (page 100)
- Herby Double Summer Bean Salad (page 110)
- Ice cream + toppings

Monday—Fun Fried Night
- The Corn Salad I Think About All Year (page 113)
- Summer Fritto Misto (page 97)
- More ice cream

Tuesday—Maybe go out?

Wednesday—Hot Dog Night, with lots of condiments and
- A Chic! Potato Salad (page 117)
- Tie-Dye Blondie-Brownies (page 183)

Thursday—Pasta Night
- Clam & Corn Pasta (page 141)
- Lemon Granita & Cream (page 147)

Friday—Brunch! Go out for dinner so you don't have to deal with more dishes in the a.m.
- Baked Challah French Toast (page 127)
- Bacon and/or eggs
- The Bloody Mary Bar (page 130)

Saturday—Leave (sadly), hopefully after eating leftover French toast.

Once the plan is in place, make lists. Three of them:
- **Food Packing List**—food you're bringing with you
- **Supply Packing List**—supplies you're bringing with you
- **Shopping List**—what you need to buy once you arrive

Lastly, if this is a group trip, dinner should be a group effort. Maybe someone's a grill goddess; someone else always knows where to shop for good produce; someone else is highly organized and will make the grocery list. Share the load with friends, and it starts to feel less like work and more like fun.

RANCH-ON-EVERYTHING DIP

Makes about 1½ cups

Zest of 1 lemon

Juice of ½ lemon

2 garlic cloves, grated

½ bunch dill, leaves and tender stems chopped

½ cup mayonnaise

½ cup sour cream

2 tablespoons onion powder

2 to 4 tablespoons buttermilk

Kosher salt and freshly ground black pepper

NOTE: *If you'd like to make this dip in advance, replace the fresh garlic with 2 teaspoons garlic powder and store in an airtight container in the refrigerator for up to 2 days.*

If dips were people, ranch would be the one everyone wants at their party. Specifically, *this* ranch, which strikes an ideal balance between fresh and familiar. Yes, you've got that richness from the sour cream and mayonnaise you know and love—but you also get a dill-y, lemony burst that makes you feel like you're frolicking in a (Hidden) valley. If you have other tender herbs in the refrigerator (like tarragon, basil, or mint), feel free to add them in—the more the merrier.

In a medium bowl, combine the lemon zest, lemon juice, garlic, dill, mayonnaise, sour cream, onion powder, 2 tablespoons of the buttermilk, ½ teaspoon salt, and lots of pepper.

If you'd prefer more of a dressing consistency than a dip, thin with more buttermilk. Taste and add more salt or lemon juice as needed. Serve with raw or roasted vegetables, any kind of salad, as a sandwich spread, or with (delivery or homemade) pizza for dipping.

season to (your) taste

I REMEMBER STANDING IN THE KITCHEN of my first apartment in Brooklyn, staring at a page in a cookbook. I had arrived in New York only a few weeks prior, fresh off the plane from college to start my first big, scary, adult job. I (and my wallet) had quickly realized I was going to have to figure out how to cook. Like, *really* cook. My best friend and college roommate, Skyler, had taught me everything I knew up until that point, which included classics such as "Meaty Pasta" (1 pound beef + 1 pound pasta + 1 jar sauce = dinner) and Trader Joe's frozen dumplings. But that's about as far as I had gotten.

So I bought a couple of cookbooks (Marcella Hazan's *Essentials of Classic Italian Cooking,* Ottolenghi's *Jerusalem*), and I would come home from work, exhausted after staring at spreadsheets, and turn my attention to their pages. It was exactly what I needed, and not just because I needed to eat. But one phrase kept coming up across cookbook pages and recipes online: "Season to taste." The recipes would often end there, abruptly, as if deciding they were tired of instructing me and were ready for me to figure it out, alone, please!

I mostly understood "Season." Seasoning is salt and pepper, right? Okay, at least I could start there. But "to taste" is where I really got stuck. It felt like a riddle. It felt like everyone else knew about this omniscient, all-knowing arbiter of "taste," except for me. (My therapist would probably have some things to say here about my issues with authority figures.) But *whose* taste?!

The moment I became a better, more confident cook, is the moment that it clicked for me: This canonical phrase is missing a word. It's missing *your*. It really should be: Season to **your** taste.

It turns out, that omniscient character I was looking for in my confusion was actually me. And it's you, too. You're not a robot following a recipe! You're the chef. And if you don't feel like one yet, start by learning how to season to *your* taste.

As you cook your way through the recipes in this book, or any recipes in any cookbook, taste them every step of the way, but *especially* right before you serve them. Decide for yourself if the dish tastes exactly how you want it, or if it's missing a little something. Here are the four ingredients I most often use to season to (my) taste:

SALT · · · · · · · · · · · · · · · · · · ·

Salt isn't meant to make your food taste salty. It's meant to make your food taste more delicious, more like itself. Salt enhances pretty much any other ingredient it touches, and as such, it's one of your most powerful cooking tools. Add a little bit at a time and watch how just an extra pinch or two can take a dish from great to amazing.

PEPPER • • • • • • • • • • • • • • •

Freshly ground, please. Cracking the pepper the moment you're seasoning gives the flavor a much more alive, dynamic taste than anything you'll pour out of a spout. If your dish is ever feeling just a touch one-note, or it's craving some earthy brightness, or you're looking for something to balance a richer, cheesier dish, crack in some pepper and see what that does for you.

LEMON • • • • • • • • • • • • • • •

I don't feel safe at home unless I have a few lemons sitting in my refrigerator. Just one spritz of lemon can be a jolt to the system—the thing that takes a salad or a salsa from nice to WOW. And while I always season with lemon to my own taste, before serving, I also love to let my guests do their own spritzing. It's never a bad idea to cut up a few wedges, place them in little bowls on the table, and let everyone go to town.

OLIVE OIL • • • • • • • • • • • • •

More specifically, good olive oil (see page 15), for a last hit of richness, fattiness, and/or pepperiness right before serving. I typically season all my pasta bowls with one last drizzle of oil, and the same goes for my salads, roasted vegetables, main-dish proteins, and more.

LE GRAND SHRIMP SALAD

Serves 6 to 8

Imagine you get a phone call: (Your version of) the Queen is coming over. For lunch. This afternoon. What will you serve?!? If I received a phone call saying the Queen—or Nora Ephron, Diane Keaton, Joan Didion, Beyoncé, my grandmother, or any of the other most important and iconic people in my life—was coming over for lunch, I would make this dish. Le grand aioli is a traditional French recipe that—in an ever-so-French way—is more like an idea, not of *what* to eat for lunch, but rather, *how* to eat it. Le grand aioli, does have, yes, aioli—but more than that, it's an excuse to gather a baby bassinet's worth of produce on a big, beautiful platter and go to town dipping. Anything can be le grand, and this shrimp salad is proof.

FOR THE SPICY-CREAMY SHRIMP SAUCE

Zest of 1 lemon

Juice of ½ lemon, plus more if needed

¼ cup cornichons, chopped

1 cup mayonnaise

2 tablespoons Dijon mustard

2 tablespoons hot sauce (I like Tabasco), plus more to taste

2 tablespoons prepared horseradish

1 tablespoon hot smoked paprika

1 tablespoon garlic powder

2 tablespoons minced fresh chives, plus more for serving (optional)

Kosher salt and freshly ground black pepper

FOR THE SHRIMP SALAD

1 lemon

½ cup kosher salt

¼ cup sugar

1½ to 2 pounds jumbo shrimp (preferably 16/20s), peeled and deveined (see Note)

2 fennel bulbs, or 6 celery stalks, thinly sliced

1 bunch scallions, thinly sliced

Freshly ground black pepper

Hot sauce

FOR SERVING

1 large head radicchio and/or butter lettuce leaves

Raw radishes, snap peas, Persian cucumbers, cherry tomatoes, and/or celery, whole or sliced lengthwise

Blanched asparagus or green beans (optional)

Jammy eggs (optional)

Ritz crackers

MAKE THE SAUCE: In a medium bowl, combine the lemon zest, lemon juice, cornichons, mayonnaise, mustard, hot sauce, horseradish, paprika, garlic powder, and chives (if using) and combine. Season with salt, pepper, and more hot sauce and/or lemon juice to taste. Cover with plastic wrap and refrigerate until you're ready to serve. *The sauce can be stored in an airtight container in the fridge for up to 3 days.*

MEANWHILE, MAKE THE SHRIMP SALAD: Fill a large pot with 10 cups of water. Halve the lemon and squeeze the juice into the water, then add the spent halves. Stir in the salt and sugar. Cover the pot and bring to a boil over medium-high heat. Fill a large bowl with cold water and a few cups of ice and set aside.

When the water is boiling, stir in the shrimp and turn off the heat. Cook the shrimp, uncovered, until bright pink and opaque, about

→

MAKE-AHEAD:

Sauce—up to 3 days

Shrimp salad—up to 4 hours

3 minutes. Use a slotted spoon or spider to transfer the shrimp to the ice bath to cool for 10 minutes. Drain the shrimp, then pat dry. Roughly chop.

Dry the bowl you used for the ice bath and place the chopped shrimp in it, along with the fennel and scallions. Add half the sauce, toss to combine, then season with salt, pepper, and hot sauce to taste. Transfer to a serving bowl or platter. *The shrimp salad is ready to go now, but you can cover and chill for up to 4 hours before serving, if desired.*

Pour the remaining sauce into a small bowl or dish, spritz it with a little lemon juice, stir, and top with some minced chives if you're feeling fancy.

Now gather your dippables: Season your lettuce leaves with salt and pile on the shrimp salad (or their own) platter. Add raw and/or blanched vegetables alongside, seasoning with a sprinkle of salt. If including jammy eggs, slice them lengthwise, season with salt and pepper, and nestle them into the platter. Add handfuls of Ritz to the platter (not optional) and serve.

SUMMER FRITTO MISTO

Serves 8 to 12 as an appetizer (without seafood), or 6 to 8 as a main (with seafood)

Neutral oil, for frying

2 medium zucchini or summer squash, sliced into ⅛-inch-thick rounds

2 large fennel bulbs, thinly sliced

1 bunch asparagus and/or green beans (12 to 16 ounces), halved

12 squash blossoms, stamens removed

1 cup fresh sage leaves

2 lemons, thinly sliced

1 pound squid (tubes, thinly sliced, and/or tentacles) and/or large tail-on shrimp (26/30 or 31/40 count), peeled and deveined (optional)

1 cup cornstarch

1 cup all-purpose flour

Kosher salt

½ teaspoon freshly ground black pepper

½ cup dry white wine, Prosecco, or water

1 cup cold seltzer

Prosecco, for serving

A Quick Aioli (recipe follows), for serving

Fritto misto is an excellent party trick. Imagine you walk into someone's home, the smell of something frying hits you immediately, and someone hands you a golden-crispy green bean. Could anything be better? No, it could not. So let's make you that someone.

I love to serve "fri-mi" as a casual appetizer, with people gathering around the kitchen eating fried delicacies as a before-dinner moment. In that situation, the seafood is entirely optional, and you could just focus on summer produce (squash blossoms are especially worth seeking out). Another route is to serve this dish as a main course by adding squid and/or shrimp to the fritto mix, perhaps alongside a fresh summery salad. In any case, one part of this recipe is nonnegotiable: taking a pause from your frying endeavors to enjoy the first batch, warm and fresh, with your guests—and ideally a spritz or glass of Prosecco.

Fill a large heavy-bottomed pot, such as a Dutch oven, with 4 inches of the oil and heat over medium heat until the oil temperature reaches 350°F on an instant-read thermometer.

On a baking sheet, create three piles: the zucchini, fennel, asparagus, blossoms, and sage leaves; the lemon slices; and the squid and/or shrimp (if using).

Line a second baking sheet or large plate with paper towels.

When you're ready to fry, in a medium bowl, whisk together the cornstarch, flour, ½ teaspoon salt, and lots of pepper. Whisk in the wine and seltzer until just combined. Dunk about two handfuls of the vegetables into the batter, letting the excess drip off. One or two pieces at a time, drop the battered veg into the oil. Fry until golden, 3 to 4 minutes, turning halfway through with a fine-skimmer, spider, or slotted spoon, then transfer to the prepared baking sheet to drain. Immediately season with salt. Skim any burnt bits of batter from the oil and adjust the heat as needed to keep oil at 350°F. Continue frying the vegetables.

Repeat this process with the seafood (if using), frying until golden, turning halfway through, 1 to 2 minutes.

Finally, repeat with the lemon—be careful, the slices can spatter furiously—until golden, 1 to 2 minutes.

→

PAIR WITH: *Tonnato (page 138), Not Another Burrata Recipe (page 137), Clam & Corn Pasta (page 141), Fluffy Sheet Pan Focaccia (page 220), Marinated Olives for Anytime (page 80), Side Lasagna (page 255)*

Transfer the fritto misto to a serving plate (or just move the baking sheet near your guests and start a fresh one). Serve this round of fritto misto now, while it's hot, with glasses of Prosecco for the ultimate experience.

Repeat in as many batches as needed, relining the baking sheet with fresh paper towels after each batch. Serve immediately, with more Prosecco and the aioli on the side.

- -

A Quick Aioli

Makes about ½ cup

½ cup mayonnaise

1 garlic clove, minced

½ lemon, cut into wedges

Kosher salt

In a small bowl, stir together the mayo and minced garlic. Squeeze one wedge of lemon into the bowl with a pinch of salt and stir. Season with more lemon juice and salt until it tastes perfect for dipping.

GRILLED CHICKEN SANDWICHES WITH MELTED SWISS & SLAW

Makes 8 sandwiches

5 garlic cloves, peeled

3 tablespoons apple
cider vinegar

2 tablespoons Worcestershire
sauce or soy sauce

1 tablespoon Dijon mustard,
plus more for serving

Kosher salt and freshly ground
black pepper

½ cup extra-virgin olive oil,
plus more for grilling

3 pounds boneless, skinless
chicken breasts

8 slices Swiss cheese

8 brioche or potato buns, split

Crunchy, Creamy Buttermilk
Slaw (page 103)

Crunchy, Roasty Glitter
(page 34; optional)

NOTE: No grill, no problem.
Heat a cast-iron skillet or grill
pan over medium-high heat
until smoking. Lightly oil the
pan before adding the chicken

PAIR WITH: *A Chic! Potato
Salad (page 117), Herby Double
Summer Bean Salad (page 110),
Gochugaru-Spiked Veg
(page 170)*

I have a confession: I prefer chicken breasts to chicken thighs. Any day, any time, in almost any setting, I will choose a chicken breast over a thigh, even though I know I run a very real risk of getting a piece of meat that is overcooked and dry. But here is my promise: Follow these steps, and you will not fall into the dry, overcooked trap.

It's all about the two-step combination of butterflying, then marinating. Juicy chicken, melty cheese, crunchy creamy slaw—this is summer in a sandwich. (And if I still haven't convinced you, yes, of course, you can use boneless, skinless thighs.)

Grate the garlic into a large bowl. Whisk in the vinegar, Worcestershire, mustard, 1 teaspoon salt, and 1 teaspoon pepper until smooth. Slowly whisk in the olive oil until emulsified.

Pat the chicken dry, then butterfly it: Working with one breast at a time, place the chicken breast on a cutting board and lay your hand on top of the breast. Hold your knife parallel to the cutting board and slice from the thickest part of the breast, cutting almost to the other side. Open it like a book; it will look a little like a heart. If the breast is on the smaller side and the little heart looks perfectly sandwich-size, leave it intact. If it's a larger breast, keep cutting all the way through to create two slimmer breast pieces. Place the butterflied breast directly into the marinade and repeat with the remaining chicken. Let sit at room temperature for 30 minutes or cover and refrigerate for up to 4 hours (if refrigerated, pull the chicken out of the fridge 45 minutes before grilling to allow it to come to room temperature).

Meanwhile, prepare a gas or charcoal grill for direct medium-high heat, 375° to 450°F, and lightly oil the grates. Remove the chicken from the marinade, letting the excess drip off, place on the grill, and close the lid. Grill, flipping halfway through, until cooked through (at least 160°F on an instant-read thermometer), 8 to 10 minutes. When the chicken is done, it will be firm and spring back a bit when you touch it. Transfer to a cutting board to rest for about 5 minutes.

Clean the grill grates or wipe out the skillet, then re-oil and reduce the heat to medium-low. Place slices of cheese on the bottom halves of the buns, and put them, along with the top buns, on the grill. Toast all the buns, covered, until golden on the bottom and the cheese is melted, 2 to 5 minutes. Transfer to a clean cutting board.

Place the chicken on the cheesy halves of the buns, then top each with a pile of slaw, and sprinkle with crunchy, roasty glitter, if desired. Spread the top buns with mustard, then gently smash to close. Serve immediately, with more slaw alongside.

CRUNCHY, CREAMY BUTTERMILK SLAW

Serves 10 to 12

Grilling Nights call for at least one dish that doesn't touch the fire—and ideally comes together without any cooking at all. This buttermilk ranch slaw is an ideal summer side or sandwich layer, like in the grilled chicken sandwiches on page 100. Using both green and red cabbage will make for a beautiful tie-dye slaw— but feel free to fully commit to whichever color calls to you in the moment.

1½ cups Ranch-on-Everything Dip (page 90)

1 cup buttermilk

Kosher salt

1 (3- to 4-pound) green or red cabbage, or ½ head of each, thinly sliced

1 to 2 lemons

1 small red onion, thinly sliced (see Note)

1 bunch chives, thinly sliced

Freshly ground black pepper

Crunchy, Roasty Glitter (page 34), for serving (optional)

In a medium bowl, whisk together the ranch dip, buttermilk, and ¼ teaspoon salt. Taste the dressing—it will be very rich but should also be very flavorful. If you're tasting more buttermilk than anything else, add pinches of salt until the rest of the flavors pop off your tongue, too. *The dressing can be stored in an airtight container in the refrigerator for up to 2 days.*

In the largest bowl you can find, combine the cabbage, ½ teaspoon salt, and the juice of 1 lemon. Using your hands, toss to mix well, massaging the cabbage to soften it a bit. Add the onion and chives and pour in the dressing. Toss to coat.

Season with more salt and pepper and/or lemon juice to taste. This slaw just gets better the longer it sits, so if you have time to make it earlier in the day, do it. Store the slaw in the refrigerator, and when you're ready to serve, sprinkle it with crunchy, roasty glitter (optional, but it really does make this slaw a standout).

NOTE: If you prefer a mellower onion flavor, place the slices in a small bowl and cover with cold water. Let sit for about 5 minutes, then drain before adding to the slaw.

MAKE-AHEAD: *Dressing—up to 2 days*

PAIR WITH: *Grilled Chicken Sandwiches with Melted Swiss (page 100), Braised al Pastor-ish Tacos (page 62), Slow-Roasted Shawarma-Spiced Salmon (page 75)*

CORN & STRAWBERRY POP-BY CAKE

Makes one 9-inch cake

Nonstick cooking spray

12 ounces strawberries, hulled and quartered

2/3 cup plus 2 tablespoons sugar, plus more for sprinkling

1/2 cup (1 stick) unsalted butter, at room temperature

2 large eggs, at room temperature

2/3 cup sour cream or plain full-fat Greek yogurt, at room temperature

2 teaspoons pure vanilla extract

1 1/2 cups all-purpose flour

1/2 cup medium-grind cornmeal

2 teaspoons baking powder

1 teaspoon kosher salt

1 cup fresh corn kernels

MAKE-AHEAD: *Up to 5 days*

People should pop by more often. When we think about hosting, our minds often go straight to special occasions or holidays. But what about those times when a friend texts to say, "I'm in the neighborhood. Can I come over and see you later this afternoon?" And you respond with, "Of course, yes, just please ignore the massive piles of laundry on my couch."

Rather than dropping everything you're doing to fold, think about what you've got in the house for a snack or a sweet treat. It could be as simple as crackers, cheese, and a seltzer over ice. But if you feel like whipping up a little something, my recommendation would be this cake, which is both snacky and sweet. Have it for dessert with a scoop of vanilla ice cream in the evening, enjoy it for breakfast with a dollop of yogurt and more fresh berries, or set it on the counter for a bite any time of day. There's no better way to welcome someone into your home than asking, "Want some cake?"

Position a rack in the center of the oven and preheat to 350°F. Coat a 9-inch round or square cake pan or springform pan with cooking spray. If using a round pan, line the bottom with parchment paper. If using a square pan, leave some parchment hanging over the edges.

In a small bowl, toss together 1 cup of the strawberries and 2 tablespoons of the sugar. Set aside to macerate.

In a large bowl, combine the butter and 2/3 cup of the sugar and, using a hand mixer, cream together on medium-high speed until light and fluffy, 5 to 7 minutes, scraping down the sides of the bowl as needed. With the mixer on medium speed, beat in the eggs one at a time, until creamy and combined. Beat in the sour cream and vanilla until just combined. Add the flour, cornmeal, baking powder, and salt, then beat on low until barely combined (you'll still have streaks of flour). Switch to a spatula or wooden spoon and fold in the unmacerated strawberries and the corn until just combined.

Scrape the batter into the prepared pan. Spoon the macerated strawberries and all their juices over the top of the batter, and sprinkle with sugar.

Bake, rotating the pan halfway through, until the cake is puffed, golden, and pulling away from the sides of the pan and a tester inserted into the center comes out with only a few moist crumbs attached, 60 to 70 minutes. Let cool for 20 minutes, then remove the cake from the pan and serve warm, or let cool completely, about 2 hours. *The cake can be stored, covered, at room temperature for up to 2 days or in the refrigerator for up to 5 days.*

HOT DOGS & A PEAK SUMMER PRODUCE PARTY

HOT DOGS ARE A PERFECT DINNER PARTY FOOD.
That is if, by now, you're in agreement that a "dinner party" does not have to mean we're all seated around a table eating a meal that appears in courses. If we define a dinner party as any gathering involving food with friends at home, or in a place that you treat like home, then believe me when I tell you: Hot dogs will never, ever fail you. In fact, they just might make for your (and everyone else's) favorite party of the year.

First of all: They're grilled. (Can you pan-fry a hot dog? Yes. Do I think a pan-fried hot dog is a perfect dinner party food? No.) And while a grilled steak or piece of fish is lovely, a hot dog is easy, simple, and thrillingly nostalgic. It's something everyone can look forward to. And, bonus for you as the host: It's nearly impossible to mess up.

See? What a relief. Now that we've decided on dogs, all your cooking focus can go toward everything else on the plate. This is the perfect opportunity to make the most of summer produce: the freshest corn salad, a snappy, herby green bean number, and a peach pie to impress everyone at the end of the night. For bevs: cold beer, seltzer over ice with lime, and plenty of chilled rosé and light red wines.

And don't forget your Condiment Corner. I like to have two or three types of mustard (yellow, Dijon, and spicy brown or whole-grain or maybe even honey mustard), small-diced raw onions, small-diced pickles, and an X-factor surprise no one expects (one of my favorites is Bungkus Bagus sambal goreng, a mix of spicy fried shallots and chiles). I personally love ketchup on my hot dogs, and the only acceptable brand is Heinz, always and forever.

It's a hog dog party, yes. But it's also a peak summertime party, where no one is stressed about overcooking anything, people go back for seconds and thirds, and the night stretches out for as long as everyone can keep eating and drinking. In other words: It's perfect.

Menu

For 10 to 12

**HOT DOGS WITH
CONDIMENT BAR**

**HERBY DOUBLE SUMMER
BEAN SALAD**

A CHIC! POTATO SALAD

**THE CORN SALAD I THINK
ABOUT ALL YEAR**

STONE FRUIT DREAMSICLE PIE

CHILLED ROSÉ & LIGHT RED WINE

HERBY DOUBLE SUMMER BEAN SALAD

Serves 8 to 12

2 shallots, diced

1 garlic clove, grated

Zest and juice of 1 lemon

⅓ cup white wine vinegar or champagne vinegar

2 tablespoons Dijon mustard

1 teaspoons honey

⅓ cup extra-virgin olive oil, plus more for serving

¼ cup fresh herbs, such as parsley, dill, cilantro, mint, and/or basil, chopped, plus more whole leaves for serving

Kosher salt and freshly ground black pepper

2 pounds mixed green, romano, and/or wax beans, trimmed

2 (15.5-ounce) cans white beans, such as cannellini, navy, or butter, drained, rinsed, and patted dry

Crunchy, Roasty Glitter (page 34), for serving (optional)

MAKE-AHEAD: *Up to 4 hours*

PAIR WITH: *DIY BLT Night (page 86), Grilled Chicken Sandwiches with Melted Swiss & Slaw (page 100), Slow-Roasted Shawarma-Spiced Salmon (page 75)*

I've always felt somewhat stressed when cooking in the summer. It's supposed to be all easy-breezy, right? But as soon as June hits, I'm confronted with A) a farmers' market full of so many vegetables I get a little weepy, and B) the visceral desire to spend as much time outdoors (that is, not cooking) as possible. How do I reconcile these competing feelings? With salads like this one. It requires zero action at the stove, comes together quite quickly, and is the zingy, herby, super fresh kind of dish I crave all summer long. It makes for a fantastic side, a winning potluck contribution (bonus: the flavors just keep getting better as it sits), or an entrée—just add grilled chicken or salmon and save a bit of dressing for drizzling all over.

In a large bowl, combine the shallots, garlic, lemon zest, lemon juice, vinegar, mustard, and honey. Whisk until smooth, then slowly stream in the olive oil until the dressing is emulsified. Stir in the herbs. Taste and add salt and pepper as needed.

Place the summer beans in a large zip-top bag and season with salt. Seal the bag and use the end of a rolling pin or bottle of wine to gently whack the beans until they partially smash and break open. Halve or chop the smashed summer beans, if you'd like, then place in a large serving bowl, tossing in the dressing along with the white beans. Season with more salt and pepper. Let sit for at least 30 minutes before serving (essential to let the flavors meld), or cover and refrigerate for up to 4 hours.

When ready to serve, drizzle with olive oil and a squeeze of lemon, and top with Crunchy, Roasty Glitter for some sparkle, if desired.

THE CORN SALAD I THINK ABOUT ALL YEAR

Serves 8 to 12

8 ears corn

12 ounces cherry tomatoes, halved

1 small red onion, diced

Zest and juice of 1 lime

Kosher salt and freshly ground black pepper

Pinch of sugar

2 tablespoons neutral oil

½ bunch fresh cilantro

Tomatillo-Avocado Salsa (page 61, chunky or blended; optional)

6 ounces queso fresco or feta cheese, plus more as desired

Extra-virgin olive oil, for serving

MAKE-AHEAD: *Up to 4 hours*

PAIR WITH: *DIY BLT Night (page 86), Slow-Roasted Shawarma-Spiced Salmon (page 75), Grilled Chicken Sandwiches with Melted Swiss & Slaw (page 100), Improved Mezcal Palomas (page 58)*

Yes, you can make corn salad any time of year—but it's never as good as it is in the summer, when fresh corn is at its sweetest. This is a corn salad to end all corn salads—with juicy tomatoes, salty feta, and zesty lime—especially when you dress it with Tomatillo-Avocado Salsa. But if you're short on time or just want to get back outside as quickly as possible, you can assemble the salad and finish it with just lime juice, olive oil, and salt to your liking—add two diced avocados for extra oomph.

Shuck the corn, then slice the kernels off the cobs (you should have about 4 cups). In a large bowl, combine half the kernels, the tomatoes, onion, and lime zest and juice. Toss to combine with 1 teaspoon salt, lots of pepper, and the sugar. Let sit for at least 10 minutes or up to 30 minutes.

Meanwhile, heat the oil in a large skillet over medium-high heat. When the oil is shimmering, add the remaining corn kernels, season with a big pinch of salt, and toss to coat. Cook, undisturbed, until charred on one side, 1 to 3 minutes. Toss and cook until totally charred, another 3 to 6 minutes. Transfer the corn to a plate to cool.

Reserve some whole cilantro leaves, then finely chop the remaining leaves and stems and add to the bowl with the corn-tomato mixture. Stir in the cooled charred corn and salsa (if using). Taste and add salt and pepper as needed. *The salad can be covered and stored in the refrigerator for up to 4 hours before serving.*

Stir in the cheese, then top with the reserved cilantro leaves and drizzle with the olive oil. Enjoy immediately.

STONE FRUIT DREAMSICLE PIE
with Stacey Mei Yan Fong

Makes one 10-inch pie, 8 to 12 servings

FOR THE CRUST

5 ounces graham crackers

3 tablespoons sugar

½ teaspoon kosher salt

4 tablespoons (½ stick) unsalted butter, melted

FOR THE FILLING

1 (14-ounce) can sweetened condensed milk

4 large egg yolks

2 teaspoons cornstarch

½ cup peach juice or nectar

2 teaspoons lemon zest

2 tablespoons fresh lemon juice

Pinch of kosher salt

FOR THE TOPPING

2 large or 3 small nectarines (about 10 ounces total), pitted and sliced ½ inch thick

¼ cup sugar

2 tablespoons chopped fresh mint

1 tablespoon lemon zest

2 tablespoons fresh lemon juice

Pinch of kosher salt

1 cup heavy cream

The story begins with a cold email. Stacey reached out to me a few months after I opened Big Night, asking if we happened to be hiring. I had just begun my search for the right person, and when I opened her email, I instantly knew: She was it. Big Night's first official employee is also Dolly Parton's number one fan and a talented baker, storyteller, and author of *50 Pies, 50 States: An Immigrant's Love Letter to the United States Through Pie*.

Stacey has the magic touch with a crust. I, on the other hand, have sworn off making them after too many blind-bake failures have ended up in the trash. And yet, every summer, I want to bake a pie that will impress all of my friends. Stacey created this recipe to help me (and you) do just that: no crimping or latticing required. The crunchy graham cracker crust cradles a smooth peach custard, topped with whipped cream and piled high with nectarines. It tastes like taking the first bite of summer.

MAKE THE CRUST: In a food processor, pulse the graham crackers into fine crumbs (you should have about 1 cup). Add the sugar and salt and pulse until just combined. Add the melted butter and pulse until the mixture looks like wet sand. Turn the mixture out into a 10-inch pie plate and press it over the bottom and up the sides, using either a ¼-cup measuring cup or your hands to press the crumbs down. Take your time, making sure that the crust walls are secure and the base is even. Freeze, uncovered, for at least 15 minutes or up to 12 hours.

Position a rack in the center of the oven and preheat to 375°F. Place the crust on a baking sheet and bake for 9 to 12 minutes, until it's lightly golden and smells buttery. Transfer to a wire rack and let cool completely before filling, about 20 minutes. Reduce the oven temperature to 350°F.

MAKE THE FILLING: In a large bowl, whisk together the condensed milk, egg yolks, and cornstarch until smooth and thick, 1 minute. Whisk in the peach juice, lemon zest, lemon juice, and salt.

Return the cooled crust to the baking sheet and pour in the filling. Bake for 27 to 30 minutes, until the edges are set but the center still jiggles like a soft thigh, as Stacey would say. Let the pie cool to room temperature, about 1 hour. Cover loosely and refrigerate for at least 3 hours or up to overnight to set completely.

MAKE-AHEAD: *Unbaked crust—up to 12 hours, Pie (without topping)—up to overnight*

MAKE THE TOPPING: When you're ready to serve, remove the pie from the refrigerator. In a medium bowl, combine the nectarines, sugar, mint, lemon zest, lemon juice, and salt. Let sit at room temperature for at least 15 minutes or up to 1 hour.

In another medium bowl, whip the cream with a hand mixer until stiff peaks form, 3 to 4 minutes. Using a silicone spatula, cover the chilled pie with the whipped cream. Use a slotted spoon to place the nectarine mixture all over the top, leaving the excess liquid in the bowl. Slice and serve immediately. Best eaten after at least two hot dogs.

A CHIC! POTATO SALAD

Serves 10 to 12

3 pounds baby or small Yukon Gold potatoes, scrubbed

Kosher salt

¼ cup white wine vinegar or sherry vinegar, plus more to taste

6 tablespoons good extra-virgin olive oil, plus more for serving

Freshly ground black pepper

1 bunch celery, including any leaves

1 bunch scallions, or 2 large shallots

6 hard-boiled eggs (see page 122; optional)

6 ounces cornichons

1 bunch fresh dill, parsley, or a mix (about 4 ounces)

1 lemon

¼ cup Dijon mustard

¼ cup mayonnaise (optional)

MAKE-AHEAD:

Dressed potatoes—up to 24 hours

Assembled salad—up to 2 days

PAIR WITH: *Grilled Chicken Sandwiches with Melted Swiss & Slaw (page 100), Crispiest Chicken Milanese with Spicy Balsamic Arugula (page 150)*

It took me a long time to realize that potato salads, in their creamy, mayo-laden glory, are polarizing. That's probably because 1) I love mayo, and 2) I grew up eating my mom's version of potato salad—a vinegary, refreshing, Dijon-y number, just ever so slightly kissed with creaminess. She originally got the recipe from Julia Child, and the two of us have adapted it over the years until it's become our Platonic ideal, ready to convert even those who are ambivalent or anti–potato salad. Is this a sexy dish? Maybe not. But I have to say, I do find it pretty chic.

In a large pot, combine the potatoes with ¾ cup salt. (Don't freak out about the quantity of salt. It's to properly season the potatoes—you will not be drinking the water!) Cover with water and bring to a boil over medium-high heat, then reduce the heat to maintain a simmer. Cook until the potatoes are fork-tender, 15 to 20 minutes.

Meanwhile, in a small bowl, whisk together the vinegar and 6 tablespoons of the olive oil until emulsified to make the dressing. Season with ½ teaspoon salt and and lots of pepper.

Drain the potatoes in a colander and let cool for about 10 minutes, until cool enough to handle. Halve the potatoes, or quarter them if larger, and place them in a large bowl. Drizzle with the dressing and give them a good toss to coat. Cover and refrigerate for at least 2 hours or overnight, tossing them a couple of times, if you remember.

When ready to assemble, remove the potatoes from the fridge and let them come to room temperature, about an hour (the oil may have solidified while it chilled).

Meanwhile, thinly slice the celery and scallions on the bias (if using shallots, mince them) and add them to the bowl with the potatoes. Finely chop the hard-boiled eggs, cornichons, herbs (reserving a big handful), and any celery leaves; add them to the bowl. Zest and juice the lemon into a small bowl, then whisk in the mustard and mayonnaise (if using). Pour over the potato mixture. Toss to combine, then season with more salt and lots of pepper to taste. *The assembled salad can be stored, covered, in the refrigerator for up to 2 days.*

Top with the reserved herbs and a glug of olive oil before serving.

BRUNCH WHEN YOU'RE STILL CELEBRATING THE NIGHT BEFORE

MY HUSBAND ASKED ME TO MARRY HIM inside a restaurant called Lighthouse, which had become our second New York City home. I was not surprised in the slightest. I knew exactly what was coming—at least I *thought* I knew what was coming—and the story of my making life difficult for him that day is one for another time.

What I *was* surprised by, however, was that 45 seconds after I said yes, I watched as my best friends—seventeen of them, all from different states and parts of my life—walked one by one into the restaurant. Alex had asked each of them months in advance if they would come to New York City to be there for this milestone. And after that moment? We proceeded to have a truly legendary night out celebrating.

When I woke up the next morning, my first thought was *brunch*. Regardless of how late we had been up the night before, we needed more time together, and that time needed to involve food. And we certainly weren't going to be able to walk into a restaurant as a group of nearly twenty. I don't quite know how I did it (some combination of adrenaline and pure joy), but I got myself to the grocery store, then came home and got to work. Having all my friends, who were now friends with each other, over after sharing a lazy afternoon of French toast and eggs and Bloody Marys and new inside jokes, is one of my most cherished memories.

Brunch at home with friends, in my opinion, is one of the best ways to host—whether it's a laid-back Sunday afternoon or you're celebrating something major. With just a little bit of advance prep work, this spread comes together easily and quickly (even if you're not feeling your 100% best after last night).

Menu

For 12 to 14

THE BLOODY MARY BAR

DEVILED EGGS AS YOU LIKE THEM

A BIG SMOKED FISH SPREAD

BAKED CHALLAH FRENCH TOAST

ANY-CITRUS JUICE (WITH OR WITHOUT CHAMPAGNE)

DEVILED EGGS AS YOU LIKE THEM

Serves 8 to 12

12 large eggs

½ cup mayonnaise

2 tablespoons Dijon mustard

Kosher salt and freshly ground black pepper

FOR SERVING

Sliced mild alliums, such as chives, scallions, or minced and rinsed red/sweet white onion

Chopped fresh herbs, such as parsley, cilantro, dill, tarragon

Spice blends, such as furikake, za'atar, baharat, chaat masala

Spicy condiments, such as hot sauce, chili crisp, harissa, yuzu kosho, gochujang

Fermented salty or spicy garnish, such as caviar or roe, chopped kimchi, capers, oil-packed anchovies

MAKE-AHEAD: *Cooked eggs— up to 24 hours*

Made-to-order eggs, or stovetop eggs of any kind, really, are not something I'd ever recommend when hosting more than a small group. No one wants a tepid scramble that gets colder with each additional second it sits out. Deviled eggs, on the other hand, are the ideal party egg. You can prep them entirely in advance, then put out a plethora of toppings for everyone to go wild with.

Fill a large pot with water, cover, and bring to a boil over high heat. Remove the lid and reduce the heat to medium-high. Working as quickly as you can, use a spider or slotted spoon to very gently lower the eggs into the water. Boil for 1 minute. Cover the pot, reduce the heat to low, and continue cooking the eggs for 11 minutes.

While the eggs are boiling, fill a large bowl with cold water and a few cups of ice. When the eggs are finished cooking, use the spider or slotted spoon to transfer the eggs from the hot water to the ice bath. Let cool for 15 minutes, then peel (it's easiest to peel eggs by tapping them gently all over, then removing the shell under running water). Pat the peeled eggs dry. *Prepare the deviled eggs within 2 hours or cover and refrigerate the peeled eggs for up to 24 hours.*

Halve the eggs lengthwise. Scoop out the yolks and drop them into a medium bowl and arrange the whites on a baking sheet or large plate. Smash the yolks with a fork, then stir in the mayonnaise, mustard, a big pinch of salt, and lots of pepper. Taste and add more salt and pepper as needed, then whisk the mixture until smooth (if you want it perfectly smooth, you can puree it in a food processor). You should have about 1¼ cups.

Using a spoon, fill the hollowed egg whites evenly with the yolk mixture. (Alternatively, transfer the yolk mixture to a piping bag or large zip-top bag with a small opening snipped from one corner and pipe in the filling). Set out alliums, fresh herbs, spiced blends, spicy condiments, and fermented salty or spicy garnishes in small bowls. Serve the deviled eggs and assorted toppings immediately.

6 Touches for a Luxury Bagel Spread

● Whether you purchased your bagels from a bagel shop or the grocery store, take them out of their bags and arrange them on a platter, or in a basket or bowl lined with a colorful kitchen towel or napkin. See? Already starting to feel like a party.

● Don't stop at one type of cream cheese—pick up at least two. Everyone loves options (even if those options are just whipped and regular).

● Herbs elevate everything. Add them liberally to your platter.

● Give your vegetables the star treatment. Slice, salt, and drain your tomatoes. Use a mandoline or sharp knife to slice your cucumbers super thin. Do the same with your onions, and quick-pickle those, if you think of it.

● Let people season to their own taste—make sure flaky salt, pepper, and good olive oil are readily available.

● Don't forget the lemons! Serve lemon wedges for spritzing, along with some zest in a little bowl. Smoked fish loves lemon zest.

A BIG SMOKED FISH SPREAD

Serves 8 to 12

2 to 3 large beefsteak or
  heirloom tomatoes, thinly
  sliced, or about 15 Campari
  tomatoes (about
  1 pound total)

Kosher salt

1½ pounds smoked fish, such
  as hot- or cold-smoked
  salmon or lox, trout, sable,
  or whitefish

1 medium red onion, or 1 bunch
  scallions, thinly sliced

1 medium English cucumber,
  thinly sliced

1 small bunch chives, sliced

1 bunch dill, roughly chopped

¼ cup brined capers, drained

1 lemon, cut into wedges

Assorted bagels, toasted, if
  desired, for serving

½ to 1 pound cream cheese, for
  serving (optional)

Flaky sea salt and freshly
  ground black pepper,
  for serving

Extra-virgin olive oil, for serving

You do not need to smoke your own fish to serve the best smoked fish spread. And this recipe (that isn't really a recipe) is proof. I will probably never make my own lox, and I'm okay with that—but I do know how to put together a stunning lox spread. Add a few thoughtful details to store-bought ingredients (see the box, opposite), and you can easily achieve the luxury brunch buffet of your dreams.

Line a baking sheet with parchment paper or kitchen towels and arrange the tomatoes on top. Season on both sides with a few pinches of kosher salt and let sit for at least 10 minutes or up to 1 hour.

Arrange the smoked fish on one side of a large platter or a few large dinner plates. (If using cold-smoked or lox, fold each slice gently in half; if using hot-smoked, use a fork or your fingers to flake up some of the fish, then leave the rest of the fillet intact.) Layer the onion next to the fish, followed by the cucumber. In the center, arrange the chives, dill, and capers (you can also place these smaller items in little bowls), then arrange the tomatoes and lemon wedges at the other end of the platter.

Arrange the bagels on a platter (or in a basket or bowl), and place the cream cheese in a separate small bowl alongside. Serve immediately, with a dish of flaky salt, olive oil, and a pepper grinder nearby.

BAKED CHALLAH FRENCH TOAST

Serves 8 to 12

This French toast is Big Morning canon in my house. Prep it the day before, then all you have to do the next morning is give it one final soak and pop it into the oven, at which point you're approximately 35 minutes away from custardy, rich-and-fluffy French toast bliss.

FOR THE FRENCH TOAST

Nonstick cooking spray, or
1 tablespoon unsalted
butter, at room temperature,
for greasing

1 loaf challah (about 1 pound),
cut or torn into 2-inch pieces

6 large eggs

2 cups whole milk

1 cup heavy cream

1 tablespoon pure vanilla extract

1 teaspoon ground cinnamon,
plus more as needed

¼ teaspoon freshly grated or
ground nutmeg, plus more
as needed

1 teaspoon kosher salt

6 tablespoons (¾ stick)
unsalted butter

⅔ cup packed brown sugar

2 tablespoons turbinado or
granulated sugar

Flaky sea salt

**FOR THE MAPLE WHIPPED CREAM
(OPTIONAL, BUT NOT REALLY)**

1 cup heavy cream

2 teaspoons pure vanilla extract

Pinch of kosher salt

2 to 4 teaspoons pure
maple syrup

Powdered sugar, for serving
(optional)

Fresh berries, for serving
(optional)

MAKE THE FRENCH TOAST: Position a rack in the center of the oven and preheat to 325°F. Coat a 9 × 13-inch baking dish with cooking spray.

Spread out the challah on a baking sheet and toast, tossing halfway though, until dried out and lightly golden in places, 20 to 25 minutes.

Meanwhile, in a medium bowl, beat 5 of the eggs until very well blended, about 30 seconds. Whisk in 1 cup of the milk, ¾ cup of the cream, the vanilla, cinnamon, nutmeg, and ½ teaspoon of the kosher salt.

Cut 4 tablespoons (½ stick) of the butter into ½-inch pieces. Place in a medium saucepan, along with the brown sugar and remaining ½ teaspoon salt. Cook over medium-low heat, stirring, until the butter melts and the sugar dissolves into the melted butter, 2 to 4 minutes. Stir in the remaining ¼ cup cream, then carefully pour the mixture into the bottom of the prepared baking dish, tilting the dish to spread the brown sugar sauce evenly.

Scatter the toasted challah pieces over the sauce, layering as necessary. Pour the egg mixture over the top. Cover the baking dish with plastic wrap and refrigerate for at least 2 hours or up to 24 hours.

Remove the baking dish from the refrigerator and preheat the oven to 375°F. Beat the remaining egg and whisk in the remaining 1 cup milk and a pinch each of cinnamon and nutmeg. Pour the mixture over the challah. Cut the remaining 2 tablespoons butter into small pieces and scatter over the challah. Sprinkle the turbinado sugar over the top.

Bake until the casserole is puffed and deeply golden brown, 35 to 45 minutes. Let cool for 10 minutes, then shower with a couple big pinches of flaky salt.

JUST BEFORE SERVING, MAKE THE WHIPPED CREAM: In a large bowl (chilled in the freezer beforehand, if you remember), combine the cream, vanilla, and salt and, using a large whisk or a hand mixer on medium speed, beat until medium peaks form. Gently fold in the maple syrup to taste, 1 teaspoon at a time, keeping in mind that the casserole is very sweet on its own.

Serve the French toast warm, with maple whipped cream, powdered sugar, and/or berries, if desired.

MAKE-AHEAD: *Unbaked French toast—up to 24 hours*

ANY-CITRUS JUICE

Makes about 4 cups, easily scaled up

6 to 8 large oranges, such as navel, Cara Cara, or blood oranges, halved

3 or 4 large pink or red grapefruit, halved

1 lemon or lime (optional)

Kosher salt

2 to 3 teaspoons honey or sugar (optional)

MAKE-AHEAD: *Up to 24 hours*

If juicing your own orange or grapefruit juice feels over-the-top, I understand. But if you're having people over for brunch or lunch and you're planning to serve mimosas, I promise this juice is worth the squeeze. Plus: It's a fantastic task to hand over to a willing guest.

Juice the oranges and grapefruit through a fine-mesh sieve (to catch the seeds and pulp) into a liquid measuring cup until you have 4 cups of juice. If you'd like a smidge more floral bitterness, juice the lemon or lime through the sieve as well.

Add a pinch of salt and give the juice a stir, then taste the juice. If it seems too tart, add honey or sugar by the teaspoonful until it tastes perfect. Decant the juice into a pitcher. *Refrigerate until you're ready to serve, up to 24 hours.*

THE BLOODY MARY BAR

Makes about 6½ cups base, for 8 to 10 drinks

Sure, you can always Go Out for Brunch. But I prefer to bring the party home with a Bloody Mary Bar, complete with my homemade base, multiple liquor options, and, of course, all the garnishes. Vodka, tequila, mezcal, beer, booze-free? Pickles or celery or hot peppers or olives? Bacon or anchovies? It's up to you (and your guests) to choose their own Bloody adventure.

FOR THE BLOODY MARY BASE

5 cups (40 ounces) low-sodium tomato or tomato-based vegetable juice (I like V8)

¼ cup dill pickle or cornichon brine

4 to 6 tablespoons fresh lemon juice (from 1 to 2 lemons)

2 tablespoons prepared horseradish

2 tablespoons vinegar-based hot sauce, plus more for serving (I like Tabasco or Cholula)

1 tablespoon Worcestershire sauce

½ teaspoon kosher salt, plus more to taste

½ teaspoon freshly ground black pepper, plus more as needed

MAKE THE BLOODY MARY BASE: In a large pitcher or large liquid measuring cup, stir together the tomato juice, pickle brine, 4 tablespoons of the lemon juice, the horseradish, hot sauce, Worcestershire, salt, and pepper—do not be shy with the pepper. Taste and stir in more lemon juice, hot sauce, and salt as needed, ¼ teaspoon at a time. Cover and refrigerate for at least 1 hour or up to 2 days.

WHEN YOU'RE READY TO SERVE, SET UP THE BAR: Arrange the celery, lemon and/or lime wedges, and any briny or meaty garnish in plates or bowls. Fill a large bowl or bucket with ice. Set out the Bloody Mary base, liquor, and a few ice-cold beers (if using).

For each drink, fill a tall glass with ice and tuck in a stalk of celery. Add 1 to 2 ounces liquor, then fill the glass with the Bloody Mary base (5 to 6 ounces). Leave about an inch of room to top with beer, if you like, and stir. Add any other desired toppings on toothpicks or cocktail picks. Serve immediately.

MAKE-AHEAD: *Base—up to 2 days*

FOR SERVING

Leafy celery stalks

Lemon and/or lime wedges

Briny garnish, such as pickle spears, cornichons, pickled peppers, green olives, and/ or cocktail onions (optional)

Meaty garnish, such as cooked bacon strips and/or oil-packed anchovies (optional)

Chilled vodka, tequila, or mezcal

Chilled light Mexican lager (I like Modelo, Tecate, or Pacífico; optional)

AN ITALIAN VACATION WITHOUT LEAVING YOUR HOUSE

THERE COMES A TIME IN EVERY summer when you simply can't escape it: Everyone you know is on vacation. Maybe they're on a Greek island. Maybe they're on some sandy beach in Florida. Maybe they're sipping rosé in Provence. You, on the other hand, are . . . not.

Stop scrolling. Put your phone down. Turn on a fan and let it blow through your hair. Or just stick your face in front of the AC. Then start planning your very own trip. Call up your friends or neighbors or maybe even a total stranger you just met in the grocery store and tell them you're going to Italy. Together.

Some people may have their vacations, but you have a Spritz Bar. And pasta. And burrata. And the freshest farmers' market tomatoes with tonnato. And you have it all at home, surrounded by friends, old and/or new.

Menu

For 8

ANY-AMARO SPRITZES

TONNATO WITH TOMATOES

NOT ANOTHER BURRATA RECIPE

CLAM & CORN PASTA

LEMON GRANITA & CREAM

NOT ANOTHER BURRATA RECIPE

Serves 6 to 8, easily scaled up

Seasonal produce, cured
   meats, fresh herbs, and
   nuts of your choice
   (see below)

8 ounces burrata

Vinegar or fresh lemon juice

Extra-virgin olive oil

Flaky sea salt and freshly
   ground black pepper

Chile flakes (optional)

PAIR WITH: *Practically any
other appetizer, salad, or main
dish in this book*

Good burrata cheese is one of those ingredients that asks so little of us—it's perfect on its own. Which is exactly why I'm not going to give you yet another burrata recipe. I will, however, use this space to remind you that a couple of balls of burrata, plus a couple simple toppings, makes for a never-fail starter. Here's some inspiration for serving it in every season.

Slice the seasonal produce however makes you happy. Arrange in a large shallow bowl. Pull apart the burrata and arrange on top of the produce. Drizzle the vinegar or lemon juice over the top, then nestle in and/or sprinkle over the other solid ingredients (meat, herbs, nuts). Drizzle with olive oil, then sprinkle with flaky salt, pepper, and chile flakes, if desired. Serve immediately.

· ·

4 Seasonal Burrata Dishes

- **SUMMER:** 8 ounces heirloom tomatoes, 1 medium peach (about 5 ounces), 2 tablespoons sherry vinegar or red wine vinegar, 1 cup fresh basil leaves

- **SPRING:** 8 ounces sugar snap peas, 6 ounces prosciutto or mortadella, 2 tablespoons lemon juice, 1 cup fresh mint leaves

- **FALL:** 8 ounces fresh figs; ½ cup chopped toasted walnuts, hazelnuts, or pistachios; 2 tablespoons balsamic glaze or 1 tablespoon balsamic vinegar whisked with 2 teaspoons honey; 3 oregano sprigs

- **WINTER:** 1½ pounds mixed citrus, such as tangerine or grapefruit, peeled; ½ small red onion or 1 large shallot; ¼ cup Castelvetrano olives; 2 tablespoons white or red wine vinegar; 3 thyme sprigs

TONNATO WITH TOMATOES

Serves 8 to 12

2 pounds mixed heirloom
    tomatoes, thinly sliced

1½ pounds small tomatoes,
    such as Campari or cherry
    tomatoes, halved

Kosher salt

3 to 4 lemons

1 garlic clove, peeled

1 (5- to 7-ounce) tin or jar
    oil-packed tuna

4 oil-packed anchovy fillets

½ cup mayonnaise, plus more
    as needed

2 tablespoons Dijon mustard

2 tablespoons brined
    capers, drained

Freshly ground black pepper

Extra-virgin olive oil, for serving

Flaky sea salt, for serving

Chile flakes, for serving
    (optional)

PAIR WITH: *Fritto Misto (page
97)—use the tonnato in place
of a Quick Aioli for dipping*

When the tomatoes are as good as they are in peak-tomato summer, you really could just slice them thickly; season them with flaky salt, pepper, and good olive oil; and call it a dish. But here's one way to make them even more divine: with a drizzle of tonnato, aka "tuna sauce" in Italian. But let's call it by its Italian name, because "tuna sauce" really undersells this umami-laden, rich-and-creamy, bright-and-zesty tomato pairing.

Line a plate with paper towels or a kitchen towel and arrange the tomatoes on top. Season them all over with salt and set aside to drain.

Meanwhile, zest and juice the lemons. You should have 2 tablespoons of zest and at least ¼ cup of juice. Transfer the zest and 3 tablespoons of the juice to a blender or food processor, then grate the garlic into the blender. Drain the tuna and anchovies, but don't toss the oil. Add the tuna, anchovies, mayonnaise, mustard, capers, and a few good grinds of pepper to the blender. Blend the mixture on medium speed until very smooth, a minute or less. With the blender running, slowly drizzle in 2 tablespoons of the reserved tuna or anchovy oil. Taste the tonnato and season with more lemon juice and salt as needed.

Arrange the tomatoes on a serving platter and pour over as much tonnato as you'd like (I like a lot of tonnato). Drizzle with olive oil, and sprinkle with a bit of flaky salt, more pepper, and chile flakes (if using). Serve immediately.

CLAM & CORN PASTA

Serves 6

3 pounds littleneck clams

Kosher salt

6 tablespoons extra-virgin
   olive oil

¾ cup panko breadcrumbs

¼ teaspoon red pepper flakes,
   plus more to taste

Freshly ground black pepper

4 ears corn

8 garlic cloves, peeled

2 lemons

½ cup dry white wine (I like to
   use an Italian bottle I would
   serve for dinner, like an
   Etna Bianco)

1 pound long dried pasta, such
   as linguine, spaghetti,
   or bucatini

½ to 1 teaspoon fish sauce
   (optional)

2 tablespoons unsalted butter

½ bunch fresh parsley, finely
   chopped, for serving

Freshly grated Parmesan
   cheese, for serving

Clam pasta, or linguine alle vongole, isn't a difficult dish to make—and yet it always feels like a special occasion. I love the combination of in-season corn and clams so much that I had to bring them together in this dish: a true celebration of summer in one big bowl. It'll make you feel like you're staring out at the Mediterranean Sea, even if you're sitting inside your apartment with the AC blasting.

Scrub the clams and rinse off any grit (if any are open or cracked, discard). Place the clams in a colander. Fill a large bowl with cold water and stir in a big handful of salt. Place the colander in the water and let the clams sit for 15 minutes. Drain the clams, rinse, and soak in fresh cold water (no salt this time) for another 15 minutes.

Meanwhile, heat 3 tablespoons of the olive oil in a 6- to 8-quart Dutch oven over medium heat. When the oil is shimmering, stir in the panko and red pepper flakes and season with salt and black pepper. Cook, stirring often, until deeply golden brown, 4 to 6 minutes. Transfer the toasted breadcrumbs to a shallow bowl to cool. Wipe out the pot but keep it handy.

Shuck the corn, then slice the kernels off the cobs (see Note). Set the kernels aside. Place the cobs in a separate large pot, fill with water, cover, and bring to a boil over medium-high heat.

Meanwhile, thinly slice 6 of the garlic cloves and grate the remaining 2. Zest and juice 1 lemon. Drain the clams (there should be only a little, if any, grit at the bottom of the bowl) and rinse again.

Heat the remaining 3 tablespoons oil in the Dutch oven over medium-high heat. When the oil is shimmering, add the sliced garlic and stir to coat. Add the wine, clams, and a few grinds of black pepper, then cover the pot. Cook, shaking the pot occasionally to agitate the mixture, until the clams open, 4 to 7 minutes (if any do not open during this time, cover and cook for 1 to 2 minutes more; if they don't open after that, pull them out and discard). Turn off the heat. Transfer the clams to a plate and cover lightly with foil to keep them warm, reserving everything else in the pot.

→

While the clams cook, remove the corncobs from the pot and discard. Stir in 2 tablespoons salt and the pasta. Cook until the pasta is barely al dente, about 3 minutes less than what the box directs. Reserve 2 cups of the pasta cooking water, then drain the pasta and add it to the Dutch oven.

Add the corn kernels and ½ teaspoon fish sauce (if using) to the Dutch oven and return to medium-high heat. Add 1 cup of the pasta cooking water and toss vigorously, cooking the pasta to al dente and forming a thin sauce, 3 to 4 minutes. Remove the pot from the heat, add the butter, grated garlic, and lemon zest, and toss, adding more pasta water a little at a time, until the sauce is glossy. Taste and add another ½ teaspoon fish sauce if you'd like a little more funk.

Stir half the parsley into the pasta along with the lemon juice and season with more black pepper and salt to taste. Return the clams to the pot, the easiest and best vessel for serving this dish family-style. (Alternatively, if you're feeling cheffy, plate each serving of pasta and clams in dinner bowls or on plates.) Top with the toasted breadcrumbs and remaining parsley. Slice the remaining lemon into wedges. Serve with the lemon wedges and Parmesan.

NOTE: For easier corn kernel slicing, first snap your cobs in half with your hands. Then lay one cob half horizontally on your cutting board and hold your knife at an angle as you slice the kernels off the cob.

PAIR WITH: *Summer Fritto Misto (page 97), Herby Double Summer Bean Salad (page 110), Bitter Greens & Broccoli Caesar (page 248), Big Beans & Tomatoes (page 153)*

ANY-AMARO SPRITZ

Makes 1 drink

2 ounces amaro (my favorites
   are Contratto and
   Cocchi Americano)

4 ounces very dry, good-quality
   Prosecco or other sparkling
   wine

Seltzer (I like Pellegrino)

Angostura bitters (optional)

1 strip of citrus zest (lemon,
   orange, or grapefruit),
   for garnish

Castelvetrano olives, for
   garnish (optional)

Memorize this combination: 2, 4, splash. That's amaro, bubbles, and seltzer, respectively—and now you're on your way to a lifetime of perfect spritzes. It's such a short ingredient list that all three of those ingredients really matter, as do the ice and the glassware.

For the amaro, when I say "any," I mean it. If you're not sure where to start, try Aperol for something on the sweet side, Campari for something sweet-bitter, Cynar for something bitter, or Fernet-Branca for something very bitter. For sparkling wine, I implore you to buy a bottle you'd want to drink on its own— the better the bubbles, the tastier the spritz. For the seltzer, I personally prefer something with smaller bubbles, like Pellegrino, but technically any seltzer will do. Use a glass that can fit a lot of ice. Then add a few cubes more than you think you need.

Fill a chilled wine glass or rocks glass with ice. Add the amaro, Prosecco, a splash of seltzer, and a couple dashes of bitters (if using) to the glass and stir. Pop the zest and/or olives into the cocktail. Drink immediately.

The Spritz Bar

The best, most sparkling kind of DIY moment to get pretty much any party started. At this point, people are pretty familiar with the Aperol Spritz. But what about a Cappeletti Spritz? A Lillet Spritz? An Averna Spritz? So many aperitifs make for absolutely delicious spritzes, and the potential of a Spritz Bar is limitless. Here's what you need to set one up:

- 2 to 4 bottles of different amari (these are your spritz bases)

- Several bottles of quality Prosecco (figure about 7 spritzes per bottle)

- Plenty of good-quality seltzer

- Lots and lots of ice

- Prepped Castelvetrano olives and lemon or orange strips, for garnish

- Ideally, an ice bucket to keep the Prosecco and seltzer cold (if not, just keep them in the fridge)

LEMON GRANITA & CREAM

Serves 6 to 8

8 to 12 lemons (about
   2 pounds)

½ cup plus 2 tablespoons
   sugar

1 teaspoon kosher salt, plus
   more to taste

2½ cups filtered water

1 cup cold heavy cream

NOTE: Serve this in clear glasses
you can see through. I can't explain
the science behind this, but it really
does make the granita experience
even more blissful.

MAKE-AHEAD: *Up to 3 months*

The first time I tasted lemon granita was at an Italian restaurant
in Brooklyn called Lilia, where it comes in a tall, ice-cold glass,
stacked between layers of vanilla soft serve. The second time was
in Sicily, at the iconic Caffè Sicilia, where it's served with brioche
and cream—for breakfast. Caffè Sicilia may be a bit farther from
home, but both are transcendent experiences worth traveling
for. In the meantime, make this lemon granita, and bring that
transcendence to your own table.

Zest 4 of the lemons, or enough so you have ¼ cup of zest; reserve
the zest in an airtight container in the refrigerator until you're
ready to garnish. Halve and juice all the lemons, then strain the
juice into a medium bowl (you should have about 1½ cups). Whisk
in the sugar and salt until dissolved, then whisk in the water. Taste
and add another pinch of salt if it seems very sweet.

Strain through a fine-mesh sieve into two shallow glass, ceramic,
or metal pans (they do not need to be the same size or type).
Chill in the freezer, uncovered, for 1 hour. Remove from the freezer
and use a fork to scrape and break up the icy mixture. Freeze for
another 1 to 3 hours, repeating the scraping process two more
times as it freezes, until the mixture is completely icy (somewhere
between a slushie and sorbet). *Scrape the granita into airtight
containers and freeze for up to 3 months. Remove it from the
freezer 15 to 30 minutes before you're ready to serve to let it
warm up to a scrape-able consistency.*

When you're ready to serve, beat the cream in a medium bowl
with a whisk or hand mixer until the cream holds soft peaks. Scoop
the granita into chilled glasses, top with a dollop of whipped
cream and a sprinkling of the reserved lemon zest, and serve.

FALL

BIGGER NIGHTS

CRISPIEST CHICKEN MILANESE WITH SPICY BALSAMIC ARUGULA

Serves 6 to 8

If August is the month when everyone goes away for one last gasp of summer, September is when they all come home. It's a time when I love to bring all our friends back together under one roof, and when I find myself craving comforting foods to ease back into the new-pencils energy of back-to-school season. This chicken Milanese is the definition of that energy: homey but still exciting thanks to the zippy lemon in the chicken and the spicy balsamic arugula that everyone's going to want seconds of.

FOR THE BALSAMIC DRESSING

¼ teaspoon red pepper flakes

1 garlic clove, grated

¼ cup balsamic vinegar

2 tablespoons crème fraîche or plain full-fat Greek yogurt

1 tablespoon Dijon mustard

3 tablespoons extra-virgin olive oil

½ teaspoon maple syrup or honey

Kosher salt

FOR THE MILANESE

6 boneless, skinless chicken breasts (2½ to 3 pounds)

6 garlic cloves, grated

⅓ cup fresh lemon juice

¼ cup extra-virgin olive oil

Kosher salt

4 large eggs

1 cup all-purpose flour

2 cups panko breadcrumbs

⅔ cup freshly grated Parmesan cheese

⅓ cup sesame seeds

Neutral oil, for frying

TO ASSEMBLE AND SERVE

12 to 16 ounces arugula

Kosher salt and freshly ground black pepper

Wedge of Parmesan cheese

Lemon wedges

MAKE THE DRESSING: In a jar, combine the red pepper flakes, garlic, vinegar, crème fraîche, mustard, olive oil, and maple syrup. Cover and shake until smooth, about 1 minute. Taste and add salt as needed. *The dressing can be stored in the refrigerator for up to 2 days.*

MAKE THE MILANESE: Pat the chicken dry with paper towels, then slice each breast in half horizontally to make 12 thin pieces. Place the cutlets between two pieces of parchment paper and use a heavy skillet, rolling pin, or meat mallet to pound to about ¼-inch thickness.

In a large bowl, stir together the garlic, lemon juice, and olive oil. Add the chicken to the bowl and let sit at room temperature for 30 minutes, or cover and refrigerate for up to 4 hours (pull them out of the fridge 45 minutes before frying to come to room temperature).

Transfer the cutlets to a cutting board, letting the excess marinade drip off, then season with salt on both sides. In a shallow bowl, beat the eggs. Place the flour on a large plate. On a separate large plate, mix together the panko, Parmesan, and sesame seeds. Dunk each cutlet into the flour, shaking off any excess; then into the egg, letting the excess drip off; then into the breadcrumb mixture, pressing to adhere. Place each breaded cutlet on a baking sheet or large plate.

Line a separate baking sheet or large plate with paper towels. Fill a large skillet with about ¼ inch of neutral oil and heat over medium-high heat until it shimmers. Working in batches as needed, carefully add two or three cutlets to the oil and cook until deeply golden, 3 to 5 minutes per side. Transfer to the prepared baking sheet and immediately season with salt. Repeat with the remaining cutlets, adding more oil as needed between batches.

Place the arugula in a serving bowl. Lightly season with salt, then pour on the dressing and toss to coat. Finish with some pepper and Parmesan shaved with a vegetable peeler.

Serve the Milanese immediately, with the spicy balsamic arugula alongside. Add lemon wedges for spritzing and any extra dressing for dipping on the side.

MAKE-AHEAD: *Dressing—up to 2 days*

PAIR WITH: *A Chic! Potato Salad (page 117), Herby Double Summer Bean Salad (page 110), Bitter Greens & Broccoli Caesar (page 248)*

BIG BEANS & TOMATOES

Serves 6 to 8

2 to 2½ pounds mixed
   tomatoes, such as heirloom,
   cherry, and/or Campari

Kosher salt

Pinch of sugar

⅓ cup extra-virgin olive oil

1 tablespoon sesame seeds

1 tablespoon coriander seeds,
   roughly crushed

2 teaspoons mild chile flakes
   or ½ teaspoon red
   pepper flakes

2 garlic cloves, peeled

2 tablespoons red wine vinegar

1 tablespoon dried oregano

1 (15.5-ounce) can butter
   beans, drained and rinsed

½ bunch chives,
   finely chopped

2 ounces feta cheese,
   crumbled or sliced,
   for serving

Fresh oregano leaves, for
   serving (optional)

Flaky sea salt

PAIR WITH: *Crispiest Chicken
Milanese with Spicy Balsamic
Arugula (page 150), Stuffed &
Roasted Leg of Lamb (page 49),
Slow-Roasted Shawarma Spiced
Salmon (page 75), Grilled Chicken
Sandwiches with Melted Swiss
& Slaw (page 100), Bulgogi-ish
Lettuce Wraps (page 169),
Big Calzone Night (page 157)*

This recipe is like the very first light sweater you get to put on after months of sweating through your tank tops. It's transitional cooking: the best way to take advantage of that sweet in-between season when the tomatoes are as good as they get— but everything else is beginning to feel just a little cozier. This is a quick, easy side dish for not-summer-anymore, not-quite-fall.

On a cutting board, roughly chop ½ pound of the tomatoes (larger, juicier ones are best for this) and slice the rest into wedges or halves. Transfer all the tomatoes and any juices that have accumulated on the cutting board to a large bowl. Add 1 teaspoon salt and a pinch of sugar and gently toss. Set aside to marinate.

Heat the oil in a small saucepan over medium heat. When the oil is shimmering, stir in the sesame and coriander seeds and toast, swirling the pan occasionally, until slightly darkened and fragrant, 2 to 3 minutes. Pour the seeds and oil into a small bowl, add the chile flakes, and grate in the garlic. Whisk in the vinegar, oregano, and a pinch of salt.

Add the beans, chives, and half the dressing to the tomatoes and toss to combine. Transfer the tomato mixture to a shallow bowl. Pour over the remaining dressing, then top with the feta, oregano leaves (if using), and flaky salt.

LIGHTHOUSE'S HUMMUS

with Naama & Assaf Tamir

Serves 6 to 8

2 cups dried chickpeas,
  soaked in water overnight

½ teaspoon baking soda

4 garlic cloves, peeled

1 teaspoon ground cumin

2 tablespoons plus 1 teaspoon
  kosher salt, plus more
  to taste

½ cup fresh lemon juice, plus
  more to taste

1 to 2 cups good tahini

Extra-virgin olive oil

2 tablespoons Calabrian
  chiles, or 1 tablespoon
  smoked paprika, for serving

Pita, raw vegetables, and/or
  chips, for serving (optional)

In my past life as a restaurant writer, nothing gave me more simultaneous joy and grief than discovering a place I wished I had known about sooner. Lighthouse was one of those restaurants. After my first visit, I wanted to tell everyone about it. Lighthouse is where I go for fresh, delicious, good-for-you food—with a side of true comfort and community—courtesy of Naama and Assaf Tamir, the sister-and-brother duo who own it.

Lighthouse's hummus is perhaps the very best example of this feeling in action. When Assaf decided he wanted to add hummus to the restaurant's menu, he went home to Israel to research it. He traveled to six different areas around the country to learn from chefs and home cooks, all of whom made hummus in totally different ways. (If this sounds like the makings of a TV show, that's because it should be.) Ultimately, his hummus, which has been served in the restaurant for years, is entirely his own—a combination of the techniques he learned from others, as well as his own secrets. One of them is that in the restaurant, he sprouts his chickpeas, which makes his hummus easier to digest and even more delicious. If you're intrigued enough to follow suit, see the Note for more information on how to sprout. But for hosts with less time on their hands, Naama and Assaf have adapted their recipe, just slightly, to make it easy to achieve their hummus in your own home.

After soaking overnight, drain and rinse the chickpeas in a colander.

In a large pot, combine the drained chickpeas with 8 cups water and 2 tablespoons salt and bring to a boil. Reduce the heat to maintain a simmer and partially cover the pot with a lid. Stir occasionally, skimming the foam from the chickpeas every once in a while, until the chickpeas stop producing it. When the chickpeas stop foaming, add the baking soda. After about an hour, fish out a chickpea and squeeze it between your fingers. If it turns to mush, the chickpeas are finished cooking. If not, continue cooking, checking every 5 minutes or so, until the squeeze test gives you mushed chickpea.

Remove from the heat, drain, and thoroughly rinse the chickpeas in a colander. Using your index finger and thumb, remove as many chickpea skins as possible, and discard them. Don't drive yourself to obsession with this step, but do it for as long as it's enjoyable and meditative and then you can stop. The more chickpea skins removed, the smoother, more flavorful the hummus.

Scoop out and reserve ¾ cup of the cooked chickpeas. In a food processor, combine the remaining chickpeas, garlic, cumin, 1 teaspoon salt, lemon juice, and 1 cup water and blend until smooth.

With the motor running, slowly pour the tahini into the food processor. If it's hot, humid, or cold, the tahini and chickpeas will mix differently, so you'll want to watch the consistency carefully, giving the tahini time to take effect on the texture of the hummus. You may only need 1 cup tahini, or you may use up to 2 cups. When the hummus becomes spreadable, you'll know it's ready. The hummus should be thick (you'll need a spatula to move it around) but very smooth. Taste and add more salt or lemon juice if needed. The hummus can be stored in an airtight container in the refrigerator for up to 5 days.

When you're ready to serve, use a spatula to pile and swoosh the hummus onto a serving plate or shallow bowl. Scatter the reserved chickpeas over. Drizzle with olive oil and Calabrian chiles, or sprinkle with paprika, if you like. Top with a handful of chopped parsley. Serve immediately, with pita, vegetables, and/or chips alongside.

NOTE: To try Assaf's chickpea sprouting method, add this step after you've soaked your chickpeas overnight: Drain the chickpeas and place them on a paper towel–lined baking sheet. Cover them with more paper towels. Place the sheet in a cool, dry place, and start checking the chickpeas after 24 hours for just the tiniest hint of a sprout—you don't want full green action (this process could take up to 72 hours, but they're usually ready within 48 hours). When that little nub of a sprout has appeared, it's time to cook the chickpeas. Omit the baking soda step entirely and move on with the rest of the recipe as written.

MAKE-AHEAD: *Up to 5 days*

PAIR WITH: *Party Chicken with Feta & Fennel (page 27), Stuffed & Roasted Leg of Lamb with a Mountain of Herbs (page 49), Slow-Roasted Shawarma-Spiced Salmon (page 75)*

BIG CALZONE NIGHT

Makes 2 large calzones; serves 6 to 8

While we're on the subject of restaurants that inspire me: Ops makes some of the best pizza in New York City. But if I'm being honest, I don't go to Ops for the pizza. I go for the calzone—a massive, steaming-hot, cheesy pocket, that comes to the table presliced for everyone to share. This is exactly the kind of restaurant magic I always want to re-create at home.

FOR THE SAUCE

3 tablespoons extra-virgin olive oil

4 garlic cloves, thinly sliced

1 (28-ounce) can crushed tomatoes

Kosher salt and freshly ground black pepper

FOR THE CALZONES

1 tablespoon extra-virgin olive oil, plus more for greasing

1 pound prepared pizza dough (see Note)

½ pound hot or sweet Italian sausage (vegetarian, if you'd like), casings removed

½ bunch broccoli rabe, or 1 bunch broccolini, roughly chopped

Kosher salt and freshly ground black pepper

1 scant cup whole-milk ricotta

2 cups grated or shredded low-moisture mozzarella cheese

1 tablespoon lemon zest

1 cup freshly grated Parmesan cheese, plus more for serving

1 large egg, beaten

Flaky sea salt

Sesame seeds or poppy seeds (optional)

MAKE THE SAUCE: Heat the oil in a medium pot over medium heat. When the oil is shimmering, add the garlic and tomatoes and season with a big pinch of salt and lots of pepper. Bring the mixture to a boil, then reduce the heat to low and simmer until thickened a bit and absolutely delicious, 15 to 20 minutes. Taste and add more salt and pepper as needed. Remove from the heat and cover to keep warm. *The sauce can be cooled and stored in an airtight container in the refrigerator for up to 1 week or frozen for up to 6 months. Defrost and warm it before serving.*

MAKE THE CALZONES: Lightly oil a baking sheet and your hands. Remove the pizza dough from the refrigerator, divide it in half, and place it on the prepared baking sheet, tossing the dough in the oil to lightly coat. Set aside to come to room temperature.

Position a rack in the center of the oven and preheat to 425°F.

Heat the oil in a medium skillet over medium-high heat. When the oil is shimmering, add the sausage and cook, using a wooden spoon to break it up, until browned, 4 to 6 minutes. Use a slotted spoon to transfer the sausage to a medium bowl (it's okay if it's still a bit pink in places), leaving any rendered fat in the pan. If there's no fat left behind, add 1 tablespoon olive oil to the pan.

Add the broccoli rabe to the skillet and cook over medium-high heat, undisturbed, until it's starting to char, about 2 minutes. Season with salt and pepper, toss, and cook until wilted and charred, 2 minutes more. Scrape the broccoli rabe into the bowl with the sausage and let cool for 5 minutes. Stir in the ricotta, mozzarella, lemon zest, and Parmesan.

→

On a clean work surface, use your fingers or a rolling pin to stretch one piece of the pizza dough into an 11-inch round—if you find the dough is contracting as you stretch it, lift and stretch it a bit on your knuckles. Cover one side of the dough with half the filling, leaving a 1-inch border. Gently fold the empty side of the dough over the filling to cover. Use your fingers to firmly seal the edges and mold the calzone into a half-moon shape. Roll and twist the sealed edge over itself, pressing firmly to further secure it. Carefully pick up the calzone (or use two spatulas to help) and place it on one side of the prepared pan. Repeat with the remaining dough and remaining filling.

Brush the tops of both calzones with the beaten egg, then sprinkle with flaky salt and sesame seeds (if using). Use a sharp knife or kitchen scissors to slice three small vents in the top of each calzone.

Bake, rotating the pan halfway through, until the calzones are puffed and deeply golden brown, 25 to 35 minutes. Let cool for 5 to 10 minutes before serving. While the calzones cool, rewarm the sauce over medium-low heat.

When ready to serve, transfer the calzones to a cutting board and use a serrated knife to slice each calzone into quarters. Place on a serving plate, piling pieces on top of each other as needed. Pour some of the sauce into a small bowl, top the sauce with grated Parmesan, and serve alongside the sliced calzones, replenishing the sauce as needed.

NOTE: If you can't find prepared pizza dough at your grocery store, check the frozen aisle (and defrost overnight). Alternatively, call your local pizza spot and ask if they'll sell you a couple of balls of their dough.

MAKE-AHEAD: *Sauce—up to 1 week (or freeze up to 6 months)*

PAIR WITH: *Bitter Greens & Broccoli Caesar (page 248), Side Lasagna (page 255), A Big Chopped Salad (page 162), Valentine Wedge (page 234), Big Beans & Tomatoes (page 153)*

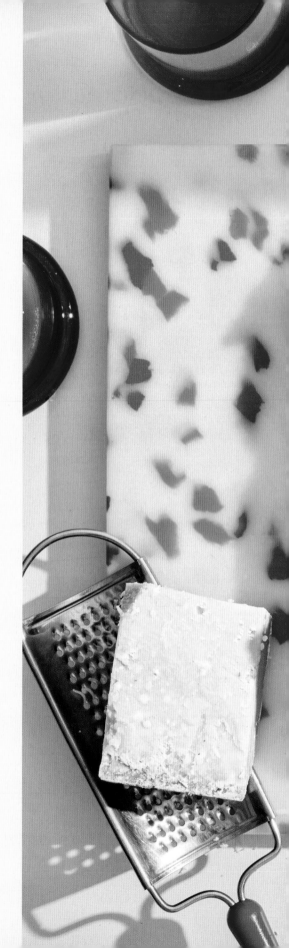

1. **Drying your lettuce is just as important as washing it.** No one wants a wet salad. If you don't have a salad spinner, gently pat the leaves dry with a paper towel or kitchen towel, and then stick them, wrapped in the towels, in the fridge. They should stay in there, dry and cold, until you are ready to assemble and toss.

rules for better salads

2. **Season every raw ingredient.** That means everything—yes, including the lettuce—you add to your salad bowl gets a sprinkling of salt, if not also a glug of olive oil and a grind of pepper, before joining the party. Tomatoes in particular should marinate in salt, olive oil, and pepper while you prep the other ingredients; drain them before adding.

3. **Bitter greens are the better salad greens.** Sure, this is a highly subjective rule, but hear me out. Radicchio is a dream to prep: Simply slice off the bottom, remove the outer leaves, and you're ready to roll with no washing and drying required. Need a salad that comes together lightning-fast? Empty out a big clamshell of prewashed arugula, and you immediately have a pile of leaves that comes with its own assertive personality, even before you add dressing.

4. To make prep that much easier, assemble and then toss your salad **in your serving vessel.**

5. That vessel matters—**you want a BIG one.** I cannot explain why salad is so much more appealing when served either in a very wide, shallow dish or a very deep bowl, but it is.

6. If you're making a leafy salad, it should be **tossed immediately before you serve it**—gently (I like to use my hands), to ensure every ingredient gets some love—lest it sit and get soggy.

7. Taste your dressing again before you toss your salad. If it tastes a little meh, try adding salt, lemon juice, and/or honey **until it tastes like YES**.

A BIG CHOPPED SALAD (TO GO WITH TAKEOUT PIZZA)

Serves 6 to 8

1 cup sun-dried tomatoes, roughly chopped, or 1 pint cherry tomatoes, halved

Kosher salt and freshly ground black pepper

½ cup extra-virgin olive oil, plus more for serving

1 small red or yellow onion, diced

2 large garlic cloves, grated

2 teaspoons dried oregano

6 tablespoons red wine vinegar, plus more as needed

2 teaspoons sugar

1 lemon, cut into wedges

1 (15.5-ounce) can chickpeas, drained and rinsed, and/or 8 ounces salami, chopped

3 or 4 heads romaine lettuce (about 3 pounds), roughly chopped

2 heads radicchio (about 1½ pounds), roughly chopped

1 cup black or green olives, such as Kalamata or Castelvetrano

Freshly grated Parmesan or Pecorino Romano cheese, for serving

Optional additions: 1 cup sliced peperoncini; 1 fennel bulb, thinly sliced; 1 cup canned or jarred artichoke hearts, drained; 2 cups fresh parsley leaves

I am a firm believer that delivery/takeout pizza night can make for a Big Night—as long as there are: friends, wine, and a big-ass salad. This could be as simple as emptying a couple of clamshells of arugula, tossing with olive oil and lemon, and grating Parmesan cheese on top—a perfect 3-minute salad. But maybe you have it in you to make a salad that just might rival the pizza itself.

I present to you: my ideal chopped salad, which could absolutely be dinner on its own but begs to share a plate with a slice of pizza. This is more template than recipe—the best chopped salads are the ones whose ingredients are exactly to your liking. Don't love olives? Leave them out. Obsessed with chickpeas? Double them. And absolutely feel free to make this vegetarian or vegan. Bonus: The only real work here is—shocker—lots of chopping. Enlist your friends to help and it will come together quick.

If using cherry tomatoes, place them in a small bowl and season with salt, pepper, and a bit of olive oil, tossing to combine. Set aside.

In another small bowl, whisk together the onion, garlic, oregano, vinegar, sugar, ½ teaspoon salt, and lots of pepper. Taste and add a squeeze of lemon juice and/or more salt as needed. Set aside at room temperature to allow the flavors to meld for at least 10 minutes or up to 2 hours. (Alternatively, cover and refrigerate for up to 2 days.)

In a large bowl, toss together the chickpeas and a glug of the olive oil and season with salt and pepper. Drain the cherry tomatoes and add them to the bowl (if using sun-dried, add them straight in).

Place the lettuce and/or radicchio in the biggest bowl you have and season with a big pinch of salt. Toss with clean hands to coat.

Slowly whisk the oil into the vinegar mixture to completely your dressing. To the bowl with chickpeas, add the olives and any other optional additions of your choosing. Pour about half the dressing over the top and toss well to combine. Pour the remaining dressing (or as much as you'd like) over the bowl of lettuce and toss to combine. Pour the contents of the large bowl into the extra-large lettuce bowl and toss with your hands until fully combined. Taste and add more salt or vinegar as needed. Grind more pepper over, shave or grate cheese over the top, and drizzle with more olive oil.

MAKE-AHEAD:
Dressing—up to 2 days

PAIR WITH: *Crispiest
Chicken Milanese
(page 150; skip the
balsamic arugula); Clam
& Corn Pasta (page 141)*

LAST-MINUTE LETTUCE WRAPS (ON A WEEKNIGHT)

"WEEKNIGHT HOSTING" HAS SUCH A NICE RING to it. And yet, I understand if those two words sound like an oxymoron or a physical impossibility. Getting through Wednesday can sometimes feel like climbing Mount Kilimanjaro all on its own—how are you supposed to shop and prep and cook and clean on top of work and meetings and errands and all your other day-to-day, just-doing-life obligations?

My best piece of advice for weeknight hosting is this: Give your friends assignments to help make dinner happen. I understand if my suggestion runs counter to your personal concept of hosting, but consider dipping your toes in the waters of dinner-prep-delegation with this Bigger Night. Here are a few ways I've asked friends for their help on this one:

- Could someone **make rice?** (If they have a rice cooker, tell them to bring the whole thing over.)
- Put someone on **lettuce duty.** Ask them to wash and separate the lettuce leaves, and bring them to the party wrapped loosely in damp paper towels in a zip-top bag.
- Our cocktails deserve fresh **ice—ask someone to grab a bag.**
- Anyone have **kimchi** (or multiple kimchis) in their fridge? Bring them over!
- Ask for a **whiskey pickup,** an **ice cream pickup,** or both.

In other words: Don't be a hero. Actually—if you're hosting on a weeknight, you already are.

And speaking of heroic: dessert. As always, vanilla ice cream dressed up a little would be more than sufficient (see page 279 for ideas). But if you want to be an overachiever, this cobbler is here for you—it's easy enough that you could actually make it once this Bigger Night has started. Just bring a friend to the kitchen with you to keep you company (read: gossip) while you do.

Menu

For 8

IDEAL WHISKEY HIGHBALLS

BULGOGI-ISH LETTUCE WRAPS

GOCHUGARU-SPIKED VEG

SAUCY SESAME SPINACH

APPLE & MISO COBBLER

PAIR WITH: *Tahini-Miso Charred Greens (page 79), Crunchy, Creamy Buttermilk Slaw (page 103), A Noodle Soup to Get People Excited (page 222)*

BULGOGI-ISH LETTUCE WRAPS

Serves 6 to 8

This is a never-fail dinner party do-it-yourself-er. While the meat is chilling in the freezer and marinating, you can use the time to pull everything else together. At a minimum, I like to have leafy lettuce, rice, ssamjang, kimchi or gochugaru-spiked veg, and herbs. When you're ready to eat, the meat cooks quickly—bring everything to the table in a DIY spread so that everyone can make their own perfect lettuce wraps.

2 to 2½ pounds boneless beef
    short ribs or rib eye steaks

5 garlic cloves, peeled

1 (1-inch) piece fresh ginger,
    peeled

½ cup unsweetened
    applesauce

¼ cup packed light
    brown sugar

3 tablespoons soy sauce

2 tablespoons gochugaru

5 tablespoons neutral oil

Kosher salt

FOR SERVING

1 bunch scallions

2 to 4 heads leafy lettuce, such
    as butter, Boston, green-/
    red-leaf, and/or radicchio

Cooked rice

Ssamjang (spicy
    soybean sauce)

Kimchi and/or Gochugaru-
    Spiked Veg (page 170)

Perilla or shiso leaves (optional)

NOTE: When cooking the meat, be careful not to crowd it in the pan. More space = crispier edges.

MAKE-AHEAD: *Marinated meat—up to 8 hours*

Wrap the meat in plastic wrap and chill in the freezer for 30 minutes (this will firm it up a bit, making it much easier to slice thin).

Meanwhile, grate the garlic and ginger into a large zip-top bag. Add the applesauce, brown sugar, soy sauce, gochugaru, and 2 tablespoons of the neutral oil.

Remove the meat from the freezer, unwrap it, and very thinly slice it against the grain (about ⅛-inch-thick pieces, if you can). Add the meat to the bag, seal it, and squish it around to mix well. Let marinate at room temperature for 30 minutes or refrigerate for up to 8 hours, tossing halfway through, if you happen to remember.

While the meat marinates, thinly slice the scallions and place in a small serving bowl. Separate the lettuce leaves from the core, wash and dry them, and place in a large serving bowl or on a platter.

When you're ready to cook, heat 1 tablespoon of the oil in a 12-inch cast-iron or nonstick skillet over medium-high heat. Remove about one-third of the meat from the marinade, letting the excess drip back into the bag. When the oil is shimmering, add the meat in a single layer. Season with a big pinch of salt and cook, undisturbed, until seared on the bottom and almost cooked through, 2 to 4 minutes. Flip, season with salt, and cook until the meat is cooked through, 1 to 2 minutes more. Transfer to a serving platter. Wipe out the skillet (you may need to deglaze it with some water first), then repeat with the remaining oil and meat in two more batches.

Set out the meat, along with the scallions, lettuce, rice, bowls of ssamjang and kimchi and/or gochugaru-spiked veg, and perilla or shiso leaves, if desired.

GOCHUGARU-SPIKED VEG

Serves 6 to 8

3 pounds crunchy vegetables, such as cucumbers, mixed radishes (red, watermelon, daikon), fennel, and carrots, thinly sliced

Kosher salt

1 bunch scallions

3 garlic cloves, peeled

¼ cup soy sauce

¼ cup unseasoned rice vinegar

3 tablespoons gochugaru

1 teaspoon sugar

1 teaspoon fish sauce (optional)

3 tablespoons toasted sesame oil

Sesame seeds, for serving

MAKE-AHEAD: *Up to 4 hours*

PAIR WITH: *Grilled Chicken Sandwiches with Melted Swiss & Slaw (page 100), Crispiest Chicken Milanese (page 150)*

Spicy, funky, and crunchy, this no-cook side holds its own on any dinner table, especially alongside meat dishes of all kinds. Don't skip the salting-then-draining step—it's key to making this simple veg dish super flavorful and crunchy.

In a colander lined with a kitchen towel or paper towel, combine the vegetables and toss with 1 tablespoon salt. Refrigerate for at least 15 minutes or up to 30 minutes.

Meanwhile, thinly slice the scallions. Grate the garlic into a large bowl. Add the soy sauce, vinegar, gochugaru, sugar, and fish sauce (if using). Whisk to combine, then slowly add the sesame oil, continuing to whisk, until the dressing is smooth.

Pat the vegetables dry and add them to the dressing along with the scallions. Using your hands, toss to combine. Taste and add more salt or vinegar as needed. Cover and let sit at room temperature for 30 minutes before serving, or refrigerate for up to 4 hours. Sprinkle with sesame seeds before serving.

IDEAL WHISKEY HIGHBALL

Makes 1 drink

Fresh ice cubes, made with filtered water (or, in a pinch, regular ice, of course)

2 ounces whiskey, such as bourbon, rye, Scotch, or Japanese whisky

4 ounces good club soda (I like Topo Chico)

Small pinch of kosher salt

Long strip of lemon peel

PAIR WITH: *Big Night Party Mix (page 180)*

While it may sound like overkill, fresh ice, made with filtered water, will make your cocktail both prettier to look at and even better to drink (and, I believe, those two things are very closely related). Once you've gathered your ingredients, all you have to do is measure, pour, and stir—easy bartending for a crowd.

Fill a highball glass with ice. Pour in the whiskey, then the club soda. Add the salt. Stir the drink once just to combine. (The ice should not float; if it does, there's too much cocktail in your glass.)

Twist the lemon peel over the drink with your thumbs and index fingers to express the lemon oils, then rub the peel around the rim of the glass. Drop the expressed peel into the drink to garnish. Serve immediately, hopefully with Big Night Party Mix.

SAUCY SESAME SPINACH

Serves 6 to 8

¼ cup doenjang or red miso paste

¼ cup unseasoned rice vinegar

2 teaspoons sugar

¼ cup toasted sesame oil, plus more as needed

6 tablespoons neutral oil, plus more as needed

2 medium yellow onions, thinly sliced

Kosher salt

4 pounds fresh spinach

3 tablespoons toasted sesame seeds

We ask a lot from vegetable sides. We want them to taste like we're actually eating vegetables, we want them to be relatively easy to put together, and we want them to complement, rather than compete with, everything else on the table. It's a tall order, and these sesame greens are up to the challenge. Four pounds of spinach sounds like a lot, but it shrinks way down, and you'll be happy you made as much as you did. The sesame seeds are an important finishing touch—they add both visual interest and a nice crunch.

In a small bowl, whisk together the doenjang, vinegar, and sugar until smooth, then slowly whisk in the sesame oil until emulsified.

Heat 2 tablespoons of the neutral oil in a large skillet over medium heat. When the oil is shimmering, add the onions and a couple of big pinches of salt and cook until just translucent, 3 to 5 minutes. Transfer to a large bowl. Add another tablespoon of oil to the skillet, along with about a quarter of the spinach, a big pinch of salt, and a splash of water. Cook until just wilted, about 1 minute, then transfer to the bowl with the onions. Repeat with the remaining spinach.

Pour the dressing over the spinach and onions, and toss to coat. Taste and add more salt and sesame oil as needed, then toss in the sesame seeds before serving.

APPLE & MISO COBBLER

Serves 8 to 10

½ cup (1 stick) unsalted butter

1½ cups all-purpose flour

¾ cup granulated sugar, plus more for sprinkling

3 tablespoons sesame seeds

2 teaspoons baking powder

3 tablespoons white miso paste

1 cup almond milk

2 large or 3 medium sweet-tart apples, such as Honeycrisp, Pink Lady, or Granny Smith, peeled, if desired, and sliced into ½-inch-thick wedges

3 tablespoons light brown sugar

1 tablespoon fresh lemon juice

Flaky sea salt

Vanilla ice cream or whipped cream, for serving

NOTE: If you happen to have any leftovers, this also makes for an amazing breakfast cake.

This cobbler is ideal for a last-minute Big Night, and it always makes me feel like I took part in the seasonal rite-of-passage that is autumn apple picking (I did it once; that was enough for me). Feel free to riff on the fruit, though—this would be fantastic with peaches or pears, if that's what's in season for you right now.

Position a rack in the center of the oven and preheat to 350°F.

Place the butter in a 12-inch cast-iron skillet and transfer to the oven to melt the butter, 5 to 7 minutes. Remove from the oven and tilt the pan around to ensure that the surface, edges, and sides are coated in butter. Set aside.

In a medium bowl, whisk together the flour, sugar, sesame seeds, and baking powder. In another medium bowl, whisk the miso and almond milk until smooth. Whisk the almond milk mixture into the dry ingredients until the batter is combined.

Wipe out the bowl you used for the flour mixture and toss together the apples, brown sugar, and lemon juice.

Pour the batter into the prepared skillet, directly over the melted butter (don't be tempted to mix this). Arrange the apples evenly over the batter, then scatter the whole cobbler with more granulated sugar.

Bake the cobbler until puffed and deeply golden brown at the edges and a tester inserted into the center comes out with only a few moist crumbs, 35 to 45 minutes. Sprinkle with flaky salt and let cool for 10 minutes before serving warm, with ice cream or whipped cream.

THE OPEN HOUSE WAY

GROWING UP IN TEXAS TAUGHT ME everything I know about hosting. Not in terms of what to cook—although I did learn queso is a universally understood love language—but how to turn a home into an open house, where everyone is always welcome.

At the parties I went to as a kid (adult parties the kids would just tag along to), the food wasn't fancy, but the snacks were abundant. People came and went all day and into the evening, getting comfy on the couch or the floor, huddling around the TV watching a football game or maybe *Dick Clark's New Year's Rockin' Eve*. (I viscerally remember a Y2K party at which we kids thought the world was about to end while the adults didn't have a care in the world.)

Most important, these parties felt relaxed. No schedule, no expectations; the point, simply, was to gather. Now, decades later, this exact kind of party is one of my favorites to host—in large part because I do not make dinner. I make a Snack Table.

A Snack Table in my house is anchored by two or three dips. I like a couple of cold options for dipping vegetables into, plus a hot, cheesy artichoke number with crackers or crusty bread (that always goes fast). There should be something sweet but not too sweet, so people can nibble on dessert all evening. And don't forget a party mix for people to munch on all day while sipping a dead-simple cocktail they can make themselves.

Welcome to my ideal open house. Take off your shoes and hang for a while. Although I will warn you, if you stay long enough, we'll end up watching YouTube videos of old Super Bowl halftime performances instead of the game.

Menu
For 12 to 16

SPAGHETTS

BIG NIGHT PARTY MIX

VEG DIP & SPREAD OF DREAMS, WITH GREEN ONION DIP AND RANCH

ARTICHOKE DIP—FOR DINNER

SWEET-SALTY PIGS IN BLANKETS

TIE-DYE BLONDIE-BROWNIES

BIG NIGHT PARTY MIX

Makes about 10 cups

½ cup (1 stick) unsalted butter, cut into pieces

⅓ cup agave nectar, or ½ cup mild honey

2 to 3 teaspoons jarred chili crisp (I like KariKari or Fly By Jing), plus 1 tablespoon oil from the jar

6 cups Chex cereal

3 cups mixed crunchy snacks, such as sesame sticks, small pretzels, Corn Nuts, and/or cheese crackers

3 cups salted potato chips

1 cup roasted, salted or unsalted, nuts, such as peanuts, almonds, cashews, hazelnuts, and/or walnuts (or more mixed crunchy snacks)

Flaky sea salt

MAKE-AHEAD: *Up to 1 week*

PAIR WITH: *Ideal Whiskey Highball (page 172), My Perfect Martini (page 240), Classic Negroni (page 29)*

If my house were a bar, this mix would be our house bar snack. Just when you think you've found every possible use for chili crisp, I'm giving you another: It's the star of this salty, nutty, crunchy mix, just waiting to be enjoyed with a cold, crisp beer.

Position the racks in the upper and lower thirds of the oven and preheat to 250°F. Line two baking sheets with parchment paper.

In a small pot, combine the butter and agave and cook over medium-low heat, stirring occasionally, until warm and runny, 2 to 3 minutes. Whisk in the chili crisp along with the oil from the jar.

In a large bowl, combine the Chex, mixed crunchy snacks, potato chips, and roasted nuts, then pour the chili crisp mixture over the top. Use your hands (wear gloves, if you'd like) or a big spatula to toss the mixture until everything is well coated. Divide between the prepared baking sheets, drizzling over any excess liquid from the bowl.

Bake for 15 minutes, give the mixture a toss, then bake for 15 minutes more. Toss again and rotate the baking sheets from top to bottom and back to front. Continue to bake, tossing the party mix and rotating the pans every 15 minutes, until fragrant and mostly crisp, 60 to 75 minutes more (it will dry and crisp further as it cools). Immediately shower with flaky salt. Let cool for at least 15 minutes before scraping into serving bowls. *Once completely cooled, the snack mix can be stored in airtight containers at room temperature for up to 1 week.*

SPAGHETTS

Makes 1 drink, easily scaled up

1 (12-ounce) bottle Miller
High Life

1 ounce Campari, Aperol,
or Cynar

Lemon wedge, for serving

PAIR WITH: *Grilled Chicken
Sandwiches with Melted Swiss & Slaw
(page 100), DIY BLT Night (page 86),
Caviar Service (page 266)*

No, this cocktail does not involve pasta. (And maybe don't text your friends in advance about it, because they might show up disappointed to find you're not making spaghetti.) It's a cocktail involving just Campari and Miller High Life. That's it, that's the entire ingredient list, not a noodle in sight.

This recipe was born at a bar in Baltimore called Wet City Brewing, and the original featured Aperol (although I personally prefer Campari or Cynar, for something a little more bitter). While you could theoretically use any crisp, pale lager, the Champagne of Beers really is the right one for the job. Set up a Spaghett Station—with a bucket of beer on ice, a bottle of Campari and/or Cynar and/or Aperol—and let everyone pour their own.

Drink (or have someone else drink) about an ounce of beer from the bottle. Pour the Campari straight into the mouth of the bottle. Squeeze the lemon wedge into the beer, and even squish it into the bottle if you're feeling it.

TIE-DYE BLONDIE-BROWNIES

Makes two 8-inch square pans

Nonstick cooking spray

1¼ cups (2½ sticks) unsalted butter, cut into pieces

1 cup packed light brown sugar

1 cup granulated sugar

2 teaspoons kosher salt

1 teaspoon baking powder

4 large eggs

1 tablespoon pure vanilla extract

1¾ cups all-purpose flour

1 cup unsweetened cocoa powder, sifted

Flaky sea salt

MAKE-AHEAD: *Up to 3 days (or freeze up to 3 months)*

I never understood the point of blondies, honestly. Why would you want a blondie when you could have a brownie? But then I discovered blondie-brownies, and it all made sense: The blondie's higher calling is to be the yang to brownie's yin. Nutty, toasty brown butter in the blondies balances the chocolate fudginess of the brownies, and together, their tie-dye superpower is unstoppable. A sweet and ever-so-slightly savory dessert that passes almost as a snack, a plate of these is perfect for an all-day graze.

Position a rack in the center of the oven and preheat to 350°F. Coat two 8-inch square pans with nonstick spray, then line with parchment, leaving an overhang on two sides.

Melt the butter in a large skillet or medium saucepan over medium heat, letting it foam and then start to brown and smell nutty, occasionally scraping the bottom and sides with a spatula to lift any browned bits, 7 to 10 minutes (don't walk away—it will seem like nothing is happening for most of this time, but then it will turn very quickly). Pour the browned butter into a medium bowl and let cool for 5 minutes. Whisk in both sugars, the salt, and the baking powder until smooth. Beat in the eggs, one at a time, until smooth and glossy. Beat in the vanilla.

Pour half the mixture (about 1½ cups) into a separate medium bowl. Whisk 1¼ cups of the flour into one bowl of the batter and set aside. Whisk the remaining ½ cup flour and the cocoa powder into the other bowl. Dollop large spoonfuls of both batters (blondie batter will be looser) evenly around the two pans. Use a small spoon to drag the batters around, creating a swirly tie-dye effect (don't go overboard or the colors will blend; it won't change the flavor, but it'll be slightly less cute).

Bake both pans on the same rack until a tester inserted into the center comes out with a few moist crumbs attached, 22 to 27 minutes. Immediately sprinkle with flaky salt, then let cool for 15 minutes. Use the parchment overhangs to lift the blondie-brownies out of the pans and set them on a wire rack to cool completely (if you have the patience or willpower not to dive in immediately), about 20 minutes.

Slice each block of blondie-brownies into 16 even squares or 32 wedges. *Store in airtight containers at room temperature for up to 3 days or in the freezer for up to 3 months.*

GREEN ONION DIP

Serves 10 to 12

1 tablespoon extra-virgin
   olive oil

2 bunches scallions

Kosher salt

2½ cups sour cream or labneh,
   or a mix

½ cup mayonnaise

2 tablespoons onion powder or
   garlic powder

2 tablespoons fresh lemon
   juice, plus more to taste

2 teaspoons soy sauce
   or tamari

½ teaspoon sugar

Freshly ground black pepper

Hot sauce (I like Zab's)

1 bunch chives

Potato chips, for serving
   (I like Ruffles)

MAKE-AHEAD: *Up to 24 hours*

PAIR WITH: *Grilled Chicken
Sandwiches with Melted Swiss &
Slaw (page 100), Creamy Tomato
Soup (page 224)*

Onion dip is such a classic party food that you'll find it jarred in just about any grocery store. No shade to those jars, but this dip is even better when actual onions are involved. Make sure to finely chop the scallions—the goal is to get a bit into every bite. This gets better the longer it sits, so if you have time to make it ahead of when you plan to serve, I would recommend you do so.

Heat a large skillet, preferably nonstick, over medium-high heat. When the skillet begins to smoke, add the oil and swirl it around to coat the pan. Add the scallions in a single layer and cook, undisturbed, until charred on the bottom, about 2 minutes. Turn the scallions and repeat on the other side. Transfer to a cutting board, season with salt, and let cool.

In a medium bowl, whisk together the sour cream, mayonnaise, onion powder, lemon juice, soy sauce, sugar, ½ teaspoon salt, a few grinds of pepper, and several dashes of hot sauce.

Trim the roots off the charred scallions and discard, then finely chop the scallions and add to the bowl with the sour cream mixture. Thinly slice the chives and add all but a handful to the bowl. Stir to combine, then taste and add more lemon juice, hot sauce, and salt until it tastes like something you don't want to stop eating. Cover and refrigerate for at least 30 minutes or up to 24 hours (in which case, pop the reserved chives in the fridge, too) to let the flavors meld.

When you're ready to serve, scoop the dip into a serving bowl and top with the reserved chives and a couple more dashes of hot sauce. Serve with potato chips for dipping.

ARTICHOKE DIP—FOR DINNER

Serves 6 to 8 as a main, or 12 to 16 as an appetizer

1 (8-ounce) block cream cheese, at room temperature

1½ cups sour cream

1 (10-ounce) package frozen spinach, thawed and squeezed to remove as much liquid as possible

2 (14-ounce) cans water-packed artichoke hearts, drained and roughly chopped

2 (4- to 5-ounce) tins high-quality oil-packed smoked trout, skin removed, broken into pieces

3 cups shredded low-moisture mozzarella cheese

1¼ cups freshly grated Parmesan cheese

1 teaspoon red pepper flakes (optional)

Kosher salt and freshly ground black pepper

Crusty bread, toasted, or tortilla chips, for serving

Crunchy vegetables, such as carrots or radishes, for serving

NOTE: When you prep this one, I suggest aiming for similar-size pieces for the artichoke and trout so they can truly share the main stage.

Artichoke dip is impossible to ignore. When that hot dish hits the table, it's like time stops and everyone turns their attention to the bubbling, cheesy appetizer before them. Why then, should this dip with main-character-energy be relegated to an appetizer? I'm giving artichoke dip the entrée spotlight it deserves, with the help of tinned trout for a little more heft. To serve, I'd recommend plenty of crunchy vegetables and crusty bread for schmearing, plus a big, simple salad on the side.

Position a rack in the center of the oven and preheat to 375°F.

In a large bowl, combine the cream cheese and sour cream. Use a fork to break up and smash the cream cheese and encourage the mixture to combine, continuing to mush and stir until the mixture is smooth.

Add the spinach, artichoke hearts, trout, 2 cups of the mozzarella, 1 cup of the Parmesan, and the red pepper flakes (if using). Add ½ teaspoon salt and lots of black pepper. Stir to combine well, then taste and add more salt as needed. Scrape the mixture into a 12-inch cast-iron skillet (or any 2½-quart baking vessel) and top with the remaining 1 cup mozzarella and ¼ cup Parmesan.

Bake until bubbling, 20 to 25 minutes. Switch the oven to broil on high and broil until the top begins to char in spots, 3 to 6 minutes, keeping a close eye.

Let the dip cool for 10 minutes, then serve warm, with crusty bread and crunchy vegetables alongside.

PAIR WITH: *A Big Chopped Salad (page 162), Spicy Balsamic Arugula (see page 150), Hidden Treasures Salad (page 198), Creamy Tomato Soup (page 224)*

SWEET-SALTY PIGS IN BLANKETS

Makes about 32 bites, easily scaled up

You know it's a party when pigs arrive in blankets. These are a little more grown-up than the usual, but they're just as irresistible, and I will take any excuse to make them (sporting or cultural event on TV = prime pigs opportunity). If you're in a hurry, you can skip the freezer step—just know that your pigs won't be quite as puffy and flaky. Then again, when something is this tasty, I doubt anyone would notice the difference.

All-purpose flour, for dusting

1 (17.3- or 14-ounce) package all-butter frozen puff pastry, thawed (see Note)

6 tablespoons fig jam/spread or apricot preserves, plus more for serving

4 ounces extra-sharp white cheddar or Gruyère cheese, grated

Freshly ground black pepper

1 (13-ounce) package cocktail wieners (about 32), patted dry

1 large egg, beaten

Flaky sea salt

Dijon mustard, for serving

NOTE: If using a 14-ounce package of dough, your pastry box will have just one thicker dough sheet. Unfold it onto a floured piece of parchment paper, roll out as needed to be about 1/8-inch thick, then halve crosswise so you have two sheets.

PAIR WITH: *Creamy Tomato Soup (page 224), Caviar Service (page 266), The Future Is Fondue (page 269)*

Dust a piece of parchment lightly with flour. Unfold one pastry sheet on the floured piece of parchment paper, and roll out to about ⅛ inch thick, then repeat with another piece of floured parchment and the other pastry sheet. Place each sheet of parchment with pastry on a baking sheet.

Place the jam in a small bowl and melt in the microwave for 10 to 15 seconds to loosen. Use a pastry brush to spread the jam evenly over both pastry sheets, dividing it evenly. Refrigerate for at least 10 minutes.

Slice each pastry sheet in half, then into quarters, then into 16 approximately 2 by 4-inch strips. Scatter the cheese evenly over the jam-covered strips, then season with pepper. Place one weiner on the bottom edge of a dough strip and roll up the pig in its blanket. Set on one of the parchment-covered baking sheets, seam-side down. Repeat with all of the pigs and dough. If you run out of wieners before you're finished with the dough, simply roll up the remaining pastry strips to create mini rolls (a bonus for any vegetarians present!). Freeze the pigs on the parchment-lined baking sheets for 15 minutes.

Position the racks in the upper and lower thirds of the oven and preheat to 425°F.

Brush the top of each pig with the beaten egg, then sprinkle with flaky salt and some pepper. Bake until golden and starting to puff, 10 to 15 minutes. Rotate the pans from back to front and top to bottom, then reduce the oven temperature to 350°F. Continue baking until the blankets are deeply golden and puffed, 20 to 25 minutes more. Let cool slightly on the pans.

Serve warm, with dishes of mustard and more melted fig or apricot jam alongside for dipping.

THE BEST FRIENDSGIVING

I WOULD BE LYING IF I TOLD YOU I love every second of shopping and prepping and cooking and cleaning every time I have people over. Before almost every dinner party, I am confronted by a brief moment when I have to pause and ask myself:

Why did I sign up for this, again?

It helps to have an answer to that question. Sometimes I host because I want to make a recipe I've been craving. Sometimes I want to celebrate a big moment. Sometimes I have a new pair of pants I need an occasion to wear. Most important, I host because I genuinely believe that bringing people together under the same roof actually brings people *together*. New friends are made, old friends catch up, and at the end of the night, while the kitchen might look like a tornado ran through it, I have a warm, fuzzy feeling that I really only get when I've created a space for my people to sit around a table, spending time together.

When deciding to host, it's important for me to—and this is the closest I will ever get to being a yoga instructor—set an intention. To remember why I'm doing this in the first place. Especially when I'm hosting Thanksgiving.

Between the media circus that starts telling you you're unprepared when it's only November 2, the complicated (to say the least) origins of this holiday, and the specific family dynamics that tend to come into focus around this time of year, I fully support you if your reaction is . . . yeah, no thanks.

My personal *why* for Thanksgiving is simple: It's a day to focus on the people—family and chosen family—I love the most. I take extra care with the food I serve, the wine I pour, and the table I set (with nary a cornucopia in sight). I don't worry about this year's hot/trendy/viral turkey recipe—my days of

spatchcocking are long behind me. Just one (excellent) stuffing will more than suffice. I make mashed potatoes, not because it would be some Thanksgiving crime not to, but because they happen to be my partner's favorite food group.

In other words, if you're the one hosting, Thanksgiving can be exactly *what* you want to make it. But don't forget your *why*, either.

Menu

For 10

NEGRONI SBAGLIATOS (PAGE 31)

CHEESE PLATE OF DREAMS

HIDDEN TREASURES SALAD

FOREVER MASHED POTATOES

ESSENTIAL CRANBERRY SAUCE

YOU-ONLY-NEED-ONE STUFFING

NOT-TRENDY, ACTUALLY DELICIOUS TURKEY & GRAVY

PUMPKIN BASQUE CHEESECAKE AND AMARO (SEE PAGE 258)

CHEESE PLATE OF DREAMS

Serves 8 to 12

Everything I know about cheese I learned from the late, great Anne Saxelby, founder of the revolutionary Saxelby Cheese. She was a champion for small, American producers making the country's best cheeses—and we have her to thank for the sea change that made these domestic delicacies more readily available for all of us to enjoy in restaurants and at home.

I also have her to thank for the Big Night Fridge—specifically, it being stocked with the very best cheeses. Here are just a few more things I learned from Anne:

- Always take your cheese out of the fridge before you want to enjoy it. Give it at least 30 minutes to get closer to room temperature, so you can really taste the layers of flavor.

- Target around 2 ounces of cheese per person when you're hosting.

- You don't need to overdo it with the number of types of cheeses. Go for quality, not quantity, and a diversity of texture/flavor. Three cheeses is an excellent number, five is a good maximum.

- Cheese doesn't really "go bad." As it ages, it changes. Soft cheeses get softer and a little funkier. Hard cheeses intensify in flavor. It's up to you to decide when a cheese in your fridge has aged a little too much for you. In any case, you do *not* need to stress about buying your cheeses the same day you serve them.

- This one is obvious but most important: The better your cheese, the better your cheese plate. Make friends with your local cheesemonger, take their recommendations, and be open to new-to-you types.

When it comes to the actual plating of my cheese plates, I am never going to be the person who's arranging salami waterfalls or artfully scattering grapes just so. (If that brings you joy, please let your creativity run wild.) My approach is to:

1. put my cheeses on a big platter or board, then cluster piles of carby, sweet, briny, and crunchy pairings around them,

2. make sure each cheese has its own serving knife, and

3. cut into each cheese before I serve, which encourages everyone else to dig in, too. No one wants to be the first one to slice. It's one of the laws of dinner party psychology.

Cheese

PICK 3 OR 4 FIST-SIZE WEDGES, ROUNDS, OR BALLS, IDEALLY A MIX OF:

- Hard cheese, such as Parmigiano-Reggiano, Pecorino Romano, Grana Padano, or Asiago

- Semi-firm/hard cheese, such as Gouda, Gruyère, aged cheddar, or Manchego

- Blue cheese, such as Roquefort, Gorgonzola, Stilton, or Cabrales

- Soft fresh or ripe cheese, such as mozzarella, chèvre, feta, Brie, Camembert, or Boursin

Something Fruity-Sweet

PICK 2 OR 3:

- Fresh stone fruit, apples or pears, grapes, figs, Fuyu persimmon

- Dried apricots, cherries, dates, or prunes

- Fig, cherry, or apricot jam

- Hot or plain honey

Something Carby

PICK 2 OR 3:

- Crusty bread, seedy crackers, buttery crackers, taralli

More Crunch or Brine

PICK 2 OR 3:

- Roasted (salted or unsalted) almonds or cashews, Marcona almonds, shelled pistachios, Corn Nuts or quicos, cornichons, olives

Something Meaty (optional)

PICK 2 OR 3:

- Cured meat, such as prosciutto, jamón, chorizo, soppressata, or hard salami

HIDDEN TREASURES SALAD

Serves 8 to 10

1½ cups pearled farro

1¼ cups apple cider

Kosher salt

2 small shallots, or 1 small red onion, diced

2 garlic cloves, grated

½ cup apple cider vinegar, white wine vinegar, or red wine vinegar

3 tablespoons Dijon mustard

Freshly ground black pepper

½ cup extra-virgin olive oil

1 cup pitted Castelvetrano olives

½ cup shelled pistachios, chopped (optional)

4 ounces Parmesan cheese, shaved

6 endives (preferably red-and-white Belgian endive)

2 small heads radicchio or Castelfranco radicchio (about 1½ pounds)

2 large or 3 small fennel bulbs, with stalks and fronds if possible (about 1½ pounds total)

1 cup fresh mint leaves

Extra-virgin olive oil, for serving

Some salads are supporting actors; this salad is a star. It's worth your attention and stomach space on a table full of other foods—yes, even (and especially) like Thanksgiving. I like to keep the farro, olive, and cheese treasures at the bottom of the serving bowl, with the leaves layered on the top for a fun surprise every time someone scoops.

This recipe is flexible, so don't get flustered if you don't have both endives and radicchio or Castelfranco. Just one type of leaf is fine! Don't have Parm? Swap in feta or even blue cheese. If you want to skip the nuts, go for it, but I'd strongly advise you keep the olives in. And, you could even add in a sweet element, like a dried cranberry, golden raisin, or sliced pear, if that's your kind of thing.

In a large saucepan, combine the farro, 1 cup of the apple cider, 1 cup water, and 1½ teaspoons salt and bring to a boil over medium-high heat. Reduce the heat to medium-low to maintain a simmer and cook, adding more water as needed if the pot gets dry, until the farro is cooked through (it should be chewy but tender), 15 to 20 minutes. Remove the pot from the heat and let cool to room temperature, or transfer to airtight containers and refrigerate for up to 2 days.

In a small bowl, combine the shallots, garlic, and vinegar and let sit for 5 minutes. Whisk in the mustard and remaining ¼ cup cider, then season with a big pinch each of salt and pepper. Slowly whisk in the olive oil until emulsified.

Gently crush the olives with the side of your knife and place in a medium bowl. Add the pistachios (if using), the cooked farro, the cheese, and ½ cup of the dressing. Toss to coat, then taste and add more salt and pepper as needed. Set aside at room temperature for up to 30 minutes or cover and refrigerate for up to 6 hours.

Separate the endive leaves and place them in a large bowl. Slice the radicchio into wedges, cut out and discard the core, and separate the wedges into large pieces; add to the bowl. Trim the bottom of the fennel bulbs. Slice the stalks from the bulbs, then separate the delicate fronds from the stalks. Halve the bulbs lengthwise, then thinly slice. Thinly slice the stalks. Add all the sliced fennel to the bowl, reserving the fronds. Sprinkle a big pinch of salt over the endive mixture, gently tossing to season. Add the remaining dressing, toss to mix well, then taste and add more salt and pepper as needed.

MAKE-AHEAD: *Farro—up to 2 days cooked, up to 6 hours dressed*

PAIR WITH: *Slow-Roasted Shawarma-Spiced Salmon (page 75), A Scallop Snack (page 233), The Future Is Fondue (page 269)*

Transfer the farro mixture to the biggest shallow serving bowl you can find. Pile the endive mixture on top of the farro mixture. Scatter over the fennel fronds and mint, tearing any large leaves, then drizzle with olive oil. Bring the dish to the table like this, without tossing.

ESSENTIAL CRANBERRY SAUCE

Makes about 2 cups, easily scaled up

2 medium blood oranges, or
   1 medium Cara Cara or
   navel orange

¾ to 1 cup sugar

1 to 1½ pounds fresh or
   frozen cranberries

¼ teaspoon kosher salt

1 tablespoon Grand Marnier
   (optional)

MAKE-AHEAD: *Up to 3 days*

If you're serving turkey, it must be accompanied by cranberry sauce. I do not make the rules. Actually, I do—we all do, at our Thanksgivings, and this is one of mine. My cranberry sauce is more tart than it is sweet, and it's much more like a jam than a jiggle. This is usually the first dish I make when I'm prepping for Thanksgiving—something about staring at a bubbling pot of magenta berries puts me in a festive mood, plus the finished sauce gets better the longer it sits in the fridge. The next day, it plays a crucial role in my Thanksgiving-leftovers sandwich (this sauce + Brie + turkey on sourdough = the real reason I cook a turkey every year).

Zest the oranges into a large saucepan. Add ¾ cup of the sugar and use your fingertips to rub with the zest until the sugar is moist and fragrant. Peel and thickly slice the oranges, removing any seeds. Finely chop the flesh and scrape it into the saucepan, along with any juices collected on the cutting board. Add the cranberries, salt, and 1 tablespoon water. Cook over medium-high heat, stirring occasionally, until the cranberries begin to break down and the juices come to a boil, 3 to 5 minutes.

Reduce the heat to medium and simmer, stirring often and smashing the cranberries as you stir, until the sauce starts to thicken, about 10 minutes. Taste and add more sugar by the tablespoon until it's the right balance of sweet and tart. Cook until the sauce is very thick and jammy, and most of the cranberries have burst but there are still plenty of whole ones, too (I think cranberry sauce looks best when you can actually see some cranberries), 5 to 10 minutes more.

When the sauce seems just about done, stir in the Grand Marnier (if using) and cook for 30 seconds more. Remove from the heat and let cool for 30 minutes. *Store in an airtight container in the refrigerator for up to 3 days.*

Remove from the refrigerator at least 30 minutes before serving, to let it come to room temperature. Your turkey deserves better than cold cranberry sauce!

FOREVER MASHED POTATOES

Serves 8 to 10

Mashed potatoes are a love language in my house—a delicacy that deserves to be enjoyed throughout the fall and winter, and not just when accompanied by turkey. While you can absolutely use either a potato masher or ricer, the ricer route always gets me the smoothest, fluffiest (and therefore, superior) result. But if you're Team Lumps, go forth with your masher—you have my support.

4 pounds Yukon Gold potatoes, scrubbed, peeled (optional), and cut into 1-inch pieces

1 head garlic (about 14 cloves), smashed and peeled

2 cups heavy cream, plus more as needed

½ bunch thyme

1 tablespoon kosher salt, plus more to taste

Freshly ground black pepper

⅓ cup sour cream, plus more as needed

Thinly sliced fresh chives, for serving

Pats of unsalted butter, for serving

NOTE: If using a ricer, leave your cooked potatoes drained in the colander until you put them, one by one, through the ricer and into the pot.

MAKE-AHEAD: *Up to 2 hours*

In a large pot, combine the potatoes, garlic cloves, cream, thyme, salt, a few good grinds of pepper, and 2 cups water (the potatoes should be mostly covered, so add a little more water if you need to). Bring to a simmer over medium heat, then cook, stirring occasionally, until the potatoes are knife-tender, 15 to 20 minutes. Remove from the heat and discard the thyme stems. Drain the potatoes in a colander set over another large pot or bowl, reserving the cooking liquid.

Return the cooking liquid to the pot and bring to a simmer over medium heat. Cook, stirring often, until reduced by about half, 8 to 10 minutes. Pour the reduced cooking liquid into its original bowl or a liquid measuring cup; you should have about 1½ cups.

Return the potatoes (see Note) and 1 cup of the reduced cooking liquid to the pot, then add the sour cream. Using a potato masher or ricer, mash or rice the potatoes until smooth, adding more cooking liquid as needed to achieve your desired level of creaminess (if you won't be serving the mashed potatoes immediately, reserve any leftover cooking liquid). Taste and add more salt and pepper as needed. *If not serving immediately, cover the mash and let sit at room temperature for up to 2 hours. Rewarm over low heat, adding more cooking water (salty and rich) or sour cream (tangy and rich) as needed to achieve your preferred flavor and texture.*

Serve warm, with chives and pats of butter on top.

YOU-ONLY-NEED-ONE STUFFING

Serves 8 to 10

1 large (1½- to 2-pound) loaf country-style sourdough bread

½ cup (1 stick) unsalted butter, cut into pieces, plus more for greasing

7 tablespoons olive oil, plus more as needed

12 ounces sweet or hot Italian sausage, casings removed

1 bunch sage

2 medium yellow onions, chopped

6 garlic cloves, thinly sliced

4 celery stalks, chopped

1 large fennel bulb, chopped

Kosher salt and freshly ground black pepper

1 cup dry white wine

½ bunch fresh parsley, finely chopped

3 large eggs

2 to 2½ cups unsalted or low-sodium chicken stock or vegetable stock

MAKE-AHEAD: *Before the second bake—up to 24 hours*

I am one of those people for whom Thanksgiving is simply an excuse to eat stuffing—on the night itself, then for breakfast the next day with a fried egg, then in a sandwich for the pinnacle of carb-on-carb action. As such, I've spent my whole life searching for the stuffing I can count on year after year to give me everything I need in this dish and nothing I don't. And sure, if someone offers to bring a second option, they're welcome to. But when stuffing is this good, I only need one.

Preheat the oven to 250°F.

Tear the bread into 1-inch pieces and divide them between two baking sheets. Bake, tossing occasionally, until totally dried out, 60 to 90 minutes. Let cool, then transfer to the biggest bowl you have.

Position a rack in the center of the oven and increase the temperature to 375°F. Grease a 3- to 4-quart baking dish.

Heat 1 tablespoon of the olive oil in a large skillet over medium-high heat. When the oil is shimmering, add the sausage and cook, using a wooden spoon to break it up, until browned and almost cooked through, 6 to 8 minutes. Use a slotted spoon to transfer the sausage to the bowl with the bread. Reserve any rendered fat in the skillet.

Pick the leaves from the sage bunch and chop enough to get 2 tablespoons (reserve the rest for later). In the same skillet over medium-high heat, add 2 tablespoons of oil and 4 tablespoons of the butter. Use a wooden spoon to scrape up any browned bits stuck to the pan. When the butter is melted, add the onions, garlic, celery, chopped sage, and fennel. Season with ½ teaspoon salt and lots of pepper, then cook, stirring often, until the vegetables have softened, 8 to 10 minutes. Add the wine and cook until the liquid is reduced by about two-thirds, about 5 minutes. Pour the mixture into the bowl with the bread and sausage. Stir in the parsley. Taste and add more salt as needed (you want it just on the verge of too salty).

In a separate large bowl, whisk together the eggs and stock (if using a 1½-pound loaf of bread, use 2 cups of stock; if using a 2-pound loaf, use 2½ cups). Pour the egg mixture over the bread mixture and use your hands to combine well. Transfer to the prepared baking dish and dot with the remaining 4 tablespoons butter. Grease one side of a foil sheet and use it, buttered-side down, to cover the dish tightly.

Bake until cooked through, 25 to 35 minutes. *(At this point, the stuffing can sit at room temperature for up to 2 hours, or let cool, cover, and refrigerate for up to 24 hours.)* Increase the oven temperature to 425°F. Uncover the stuffing and bake until warmed through and deeply golden, 20 to 35 minutes. Let sit for 10 minutes to cool slightly and allow the flavors to meld.

While the stuffing is on its second bake, heat the remaining 4 tablespoons olive oil in a small saucepan over medium-high heat. Line a plate with paper towels. When the oil is shimmering, add about half of the reserved sage leaves to the oil and fry until curling and crisp, 5 to 10 seconds. Use a slotted spoon, fine-mesh skimmer, or spider to transfer the leaves to the paper towels to drain. Immediately sprinkle with salt. Repeat with the remaining sage leaves.

Serve the stuffing warm, topped with the fried sage.

NOT-TRENDY, ACTUALLY DELICIOUS TURKEY & GRAVY

Serves 8 to 10

FOR THE TURKEY

½ cup kosher salt

1 tablespoon freshly ground black pepper

3 tablespoons light brown sugar

1 (12- to 14-pound) turkey

1 bunch thyme, sage, or oregano

2 heads garlic, halved

1 pound shallots (6 to 8), or 6 red onions, halved crosswise (or quartered if large)

¼ cup extra-virgin olive oil or neutral oil

½ cup (1 stick) unsalted butter

FOR THE GRAVY

3¾ cups turkey drippings (supplement with chicken or turkey stock)

¼ cup dry white wine (or 2 tablespoons each white wine vinegar and chicken or turkey stock)

2 to 4 tablespoons unsalted butter

¼ cup all-purpose flour

1 tablespoon white wine vinegar or apple cider vinegar

1 tablespoon Worcestershire sauce or soy sauce

Kosher salt and freshly ground black pepper

Listen, we all had our spatchcocking era. Or maybe you were into turkey smoking. Or frying. Or buttermilk brining. But I'm done chasing the trends. The turkey trends, that is. I've reached the point in my hosting life where all I want is to make an actually delicious bird for me and all my friends. So I use the method I know will get me there every time: dry brined, slow-cooked, and minimal frills.

At the end of the day, three factors decide the most delicious turkeys:

1. The turkey itself. Buy a high-quality bird from your local butcher and you will taste the difference!

2. The dry-brining time. The longer you can let your turkey brine, uncovered in the fridge, the better.

3. The cook time. Remember your oven may be lying to you about how hot it claims to be. Start checking your turkey for doneness earlier than you may think you need to. It's always better to be safe than sorry, especially on Thanksgiving.

MAKE THE TURKEY: In a small bowl, combine the salt, pepper, and brown sugar with your hands.

Fit a wire rack into a rimmed baking sheet. Remove the giblets and neck from the turkey and place it on the rack. Pat the turkey dry all over, including both cavities. Sprinkle the salt mixture evenly all over the turkey, gently pressing to adhere. Refrigerate the dry-brined turkey, uncovered, for at least 12 hours or up to 2 days (the longer it brines, the better the bird). Remove from the fridge 1 hour before roasting.

Position a rack in the bottom third of the oven and preheat to 425°F.

Use paper towels to pat the turkey dry, but take care not to rub off the brine. Transfer the turkey to a clean baking sheet or roasting pan, breast-side up. Tuck the wings underneath the top of the breast (this prevents them from getting too dark). Stuff the turkey with herbs, half a garlic head, and half the shallots. If you'd like, tie the legs together with kitchen twine. Arrange the rest of the shallots and garlic head halves around the turkey. Drizzle the shallots with oil, then season with salt and pepper. Drizzle the turkey with oil and rub it all over the bird.

Roast the turkey, rotating the pan halfway through, until the skin is just starting to turn golden, 20 to 30 minutes. (Don't worry about getting a deep, even color here—it will brown much more as it roasts.)

Meanwhile, melt the butter in a small saucepan over medium-low heat. Remove the turkey from the oven and brush all over with the butter. Return to the oven and reduce the oven temperature to 300°F.

Continue to roast, basting the turkey and shallots with the pan juices and brushing with more melted butter every 45 minutes, rotating the pan each time, until the turkey is deeply golden brown and an instant-read thermometer plunged into the thickest part of the thigh reads at least 160°F, 2½ to 3½ hours. If the thigh is not yet at 160°F, continue roasting and taking its temperature every 15 minutes, and if at any point parts of the turkey are getting too dark, tent with foil.

Remove from the oven and let cool for 15 minutes, then carefully transfer the turkey to a cutting board, letting any drippings from the main cavity fall onto the baking sheet. Transfer the shallots and any herbs from the baking sheet to a small bowl. Pour the turkey drippings left in the baking sheet through a fine-mesh sieve into a large heatproof measuring cup or medium bowl.

Let the turkey rest for at least 30 minutes before carving.

MEANWHILE, MAKE THE GRAVY: Skim and discard the fat from the turkey drippings, if desired, supplementing with stock as needed to make 3¾ cups of liquid. Add the wine to the measuring cup.

In a medium saucepan, heat the butter (if you have any extra from basting the turkey, start with that, then see how much more you need to make 4 tablespoons) over medium heat until foamy. Add the flour and cook, whisking often (it will be clumpy), until the mixture smells nutty and darkens slightly, 2 to 4 minutes.

Whisk in the turkey dripping mixture, 1 cup at a time, and cook until bubbling and combined. Increase the heat to medium-high and bring the gravy to a simmer. Cook, whisking occasionally and reducing the heat as needed, until the gravy has reduced by about one-third and is thick enough to coat the back of a spoon, 20 to 25 minutes. Whisk in the vinegar and Worcestershire and season the gravy with salt to taste. Cover the pot and keep the gravy warm over low heat until you're ready to serve, up to 2 hours.

Carve the turkey and serve (be sure all those jammy shallots make it onto the serving plate), with the gravy alongside.

MAKE-AHEAD:

Dry-brined turkey—up to 24 hours

Gravy—up to 2 hours

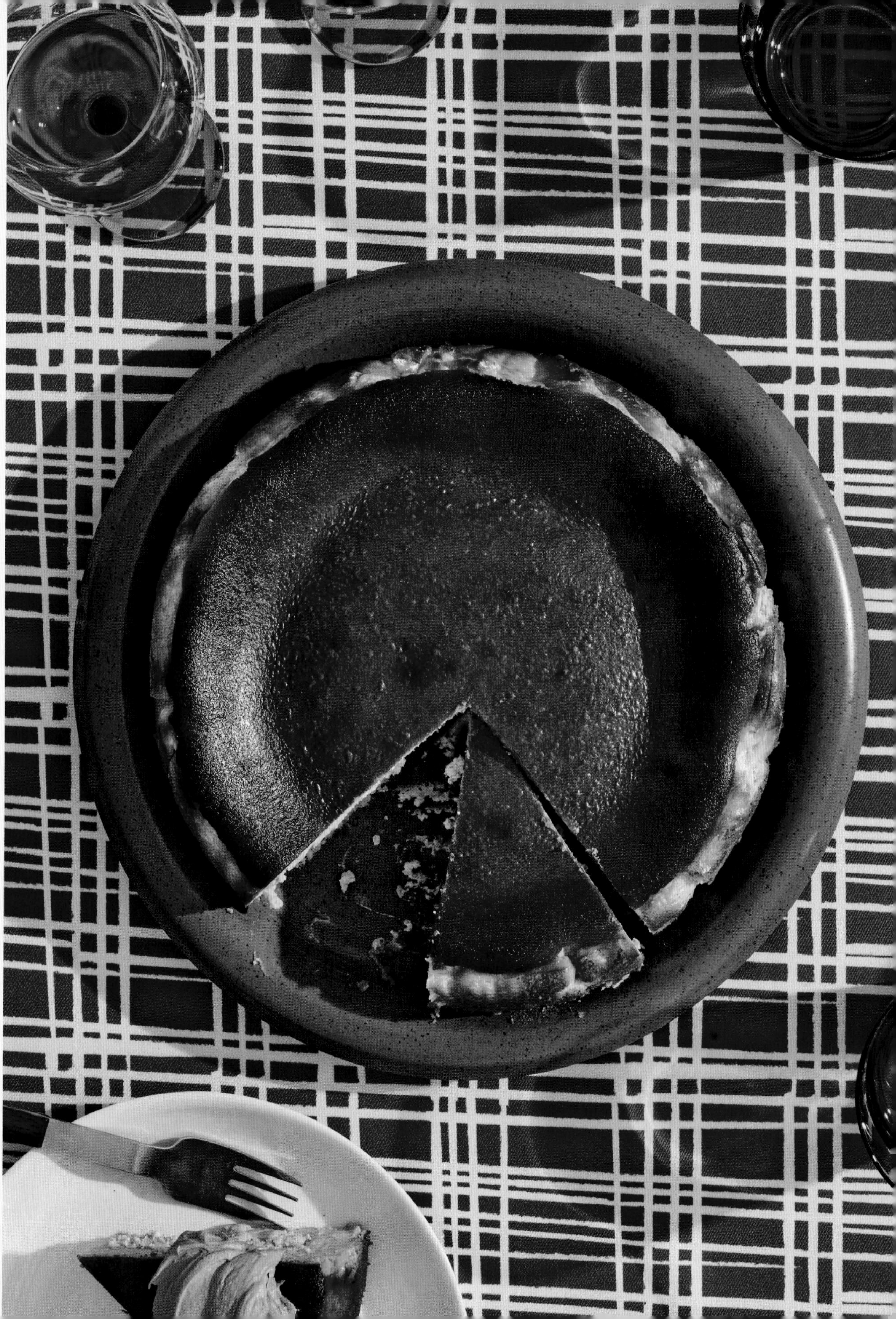

PUMPKIN BASQUE CHEESECAKE

Makes one 9-inch cheesecake

When it comes to Pie with a capital P, I generally like to leave it to the professionals. Especially around Thanksgiving—I just don't trust myself with crust, nor do I want to be the Person Who Ruined the Pumpkin Pie. Pumpkin Basque Cheesecake, on the other hand, with its almost-burnt crust, defies the need for perfection, and it is likely to please just about everybody around the table—including, thanks to the cocoa whip, the person who was really hoping for something chocolaty.

FOR THE CHEESECAKE

Nonstick cooking spray

3 (8-ounce) blocks cream cheese, at room temperature

1¼ cups granulated sugar

1½ teaspoons kosher salt

4 large eggs, at room temperature

1 cup pumpkin puree

1 cup heavy cream, at room temperature

2 teaspoons pure vanilla extract

¼ cup all-purpose flour

1½ teaspoons ground cinnamon

1 teaspoon ground ginger

½ teaspoon freshly grated or ground nutmeg

FOR THE COCOA WHIP

1 cup heavy cream

3 tablespoons unsweetened cocoa powder (preferably Dutch-process)

2 teaspoons pure vanilla extract

1 tablespoon powdered sugar, plus more to taste

Pinch of kosher salt

MAKE-AHEAD: *Cheesecake— up to 3 days (or freeze up to 3 months)*

MAKE THE CHEESECAKE: Position a rack in the center of the oven and preheat to 400°F. Place a baking sheet in the oven to preheat as well. Coat the bottom and sides of a 9-inch springform pan with nonstick spray. Tear one or two long pieces of parchment paper and overlap them in the pan, ensuring lots of overhang.

In the bowl of a stand mixer fitted with the paddle attachment, combine the cream cheese, granulated sugar, and salt and beat on medium-low speed until light and fluffy and the sugar has dissolved (rub a bit between your fingers—it should not be grainy), about 2 minutes. (Alternatively, combine the ingredients in a large bowl and, using a hand mixer, beat on low speed as directed.) Scrape down the sides of the bowl. Beat in the eggs, one at a time, until totally combined, scraping down the sides of the bowl after each addition to be sure no lumpy bits are stuck to the bottom. With the mixer on low speed, beat in the pumpkin, cream, and vanilla until combined.

Pass the flour, cinnamon, ginger, and cardamom through a fine-mesh sieve or sifter into the mixer bowl. Beat on low speed until the batter is totally smooth, about 1 minute.

Scrape the batter into the prepared pan and set the pan on the preheated baking sheet. Bake until deeply browned on top (singed in places and nearly burnt is good!) but still jiggly, 45 to 60 minutes (the timing will vary greatly based on your oven). Increase the oven temperature to 500°F and bake until even more browned, 5 to 10 minutes.

Remove from the oven and let cool completely in the pan, about 3 hours. Refrigerate the cheesecake, uncovered, for at least 8 hours or ideally overnight before unmolding and serving.

JUST BEFORE SERVING, MAKE THE COCOA WHIP:
In a large bowl, combine the cream, cocoa powder, vanilla, powdered sugar, and salt and, using a large whisk or a hand mixer on medium speed, beat until the cream holds soft peaks, 1 to 2 minutes. If you prefer it sweeter, whisk in more sugar to taste. (This will come together faster than plain whipped cream; if you accidentally overwhip, fold in more cream.)

When ready to serve, run a sharp knife under hot water, then slice the cheesecake into wedges. Serve with a dollop of cocoa whip. *Leftover cheesecake can be stored in an airtight container in the refrigerator for up to 3 days; or wrap individual slices in plastic wrap and freeze for up to 3 months.*

a trip to the wine shop
with Christine Collado

Stopping at my local wine shop is one of my most favorite pre-hosting rituals. It's a place where I always learn something new, where I feel like I'm traveling around the world (all without leaving my little corner of it), and where I've made friends with wine experts who know so much more than I do, and are generous with that knowledge.

My friend Christine Collado, a sommelier with a wildly impressive résumé that covers all areas of the wine world, is one of those experts. She worked at world-famous restaurants like Daniel and Brooklyn Fare, among other spots, before joining Parcelle Wine, a national online retailer and wine bar in Manhattan's Chinatown. She graciously agreed to share some of her best tips (in *italic* below), and yes, you have my permission to rip out this page and carry it with you on your next trip to the wine shop.

First, don't be afraid to **ask for help**, and help the staff help you by giving them some context clues:

- What's the occasion? *Is it for dinner tonight or are you going to a party? Is this a gift? The occasion is often very associated with your budget, which is also a key piece of information.*

- Speaking of, what is your budget? Have a number in mind, and let the staff know so they can help you get the best bang for your buck.

- What flavors are you into? *What do you tend to enjoy most—high acid, tart, citrus-based flavors? Or are you into things that are richer and maybe have sweeter undertones?*

- What are you planning to eat? *Or even more general—what foods do you most enjoy?*

But, and this is a big one: **Don't get too hung up on pairings**. *People tend to fall back on the pairing thing because they feel like it's the "correct" thing to do. A good way to play it safe is to trust that classic wine regions and their dishes will inherently go well together.*

If you're ready to start exploring more, think more about the details of your dinner: Maybe a spice, or a sauce, or another element of the dish—what will dinner actually taste like? Is it smoked? Is it grilled? Is it pan-seared? And what would pair well with that flavor? If you're not sure, ask!

Another way to go about it: **BYO grocery bag**, *like one of Christine's favorite customers. He would come straight from the grocery store to pick up a bottle of wine for dinner and show me what he had bought at the store. He'd say, "I have this rib eye and I think I'm going to do some type of salsa verde," and we'd go from there.*

What about wines that go well with lots of different types of foods? *Beaujolais is juicy and fresh, chill and unpretentious. It's one of* those red wines you can have for a variety of occasions, that you can pair with fish, steak, or anything grilled. White Burgundy is another flexible option, that can offer an amazing array of styles, even though it's one grape.

When you're having a dinner party:

- *Allocate one bottle per person.* You may not open them all, but you should have them at the ready.

- *If you can find an affordable large-format wine, get it.* When Christine says "large-format," she's talking about magnums: two bottles in one. You'll know it when you see it—they look like an instant party, because they are.

- Think about how you want to course your wine. *I like to have a bottle open to welcome someone—something crushable and delicious and easy—and then build from there. Maybe grab one or two special bottles that you might want to spend a little more money on than the others. Maybe add another bottle that's fresh and briny and delicious, and another fun, funky bottle that you're really into. It's all about diversifying your options.*

When you're hosting a big event with a lot of different types of people, such as, say, Thanksgiving: *It's not really about you. The most hospitable thing is to think about what would be most crowd-pleasing. It's about creating those moments where people are like, "Oh, this is really good."* Maybe your guests took a trip to South America, so you decide to pick up a pinot noir from Patagonia—something fun and interesting and a little different while still being relevant to their interests.

Bubbles aren't just for toasting. *Sparkling wines have a place at the table beyond just an aperitif. They're worthy of so much more drinking than just on special occasions.*

Wine is one of the best ways to learn about a place without leaving your house. *If you want to drink something new, and you don't know where to start, think Vacation Destinations. People love to travel to special locations that just so happen to have incredible wine cultures. The boom of Sicily and the Mediterranean is just one example.*

If your friends are asking what kind of wine to bring over to your house, tell them what you already have and encourage them to fill in the gaps. *I've got two bottles of white and a red. If you want to bring an orange wine or a bottle of rosé to start, that sounds great.*

Above all else, for best success, find your own personal wine guide. *Once you find that person or that wine shop or that blog or whatever other go-to reference makes you feel empowered—that's where you can really start developing and having a lot of fun and taking a little bit more risk with your wine choices.*

· ·

BIGGER NIGHTS

· ·

AMATRICIANA FOR A CROWD

Serves 8 to 10 as a main

1 tablespoon extra-virgin olive
  oil, plus more for drizzling

12 ounces guanciale (or
  pancetta, in a pinch),
  chopped into ¼-inch pieces

1 teaspoon freshly ground
  black pepper, plus more
  as needed

2 small red onions, or 2 to
  3 shallots, diced

1½ teaspoons red pepper
  flakes

6 garlic cloves, crushed

¼ cup dry white wine

2 (28-ounce) cans whole
  peeled tomatoes, crushed
  by hand

Kosher salt

1½ pounds short tubular pasta,
  such as rigatoni or lumache

1 cup freshly grated Pecorino
  Romano or Parmesan
  cheese, plus more for serving

NOTE: You can use
pancetta here, but I strongly
recommend guanciale for
its intensely pork-y flavor,
which stands out against
the glossy sauce.

PAIR WITH: *Bitter Greens &
Broccoli Caesar (page 248),
A Big Chopped Salad
(page 162), Not Another
Burrata Recipe (page 137)*

Amatriciana is one of the core four Roman pastas—and it's the
one that has a piece of my heart. It's a perfect sauce, and unlike
carbonara or cacio e pepe, it scales up beautifully to feed a crowd.
On that note: We're using short, tubular pasta here—rather than the
traditional bucatini noodle, which I do love but is nearly impossible
to cook evenly at scale. (Nothing is worse than a noodle that's
half overcooked and half crunchy in the middle.) Go for rigatoni or
something similar and stir it continuously as it cooks.

Heat the oil in a large, ideally high-sided pot over medium-low
heat. When the oil is shimmering, add the guanciale and black
pepper. Cook, stirring occasionally and letting the meat brown
slowly, until much of the fat has rendered and the guanciale is
very crispy, 15 to 20 minutes. Use a slotted spoon to transfer the
guanciale to a plate or small bowl.

Add the onions and red pepper flakes to the skillet and cook,
stirring often, until just beginning to soften, 3 to 5 minutes. Add the
garlic and cook, stirring, until very fragrant and starting to take on
a bit of color, about 90 seconds. Add the wine, increase the heat
to medium, and stir, scraping up any browned bits from the bottom
of the pan. Let sizzle until nearly all the wine has reduced, 2 to 4
minutes. Add the tomatoes and 1 teaspoon salt. Bring the sauce to
an active simmer, then reduce the heat to medium-low and cook,
stirring occasionally, until thickened, 25 to 30 minutes. Taste and
add more salt and black pepper as needed.

Meanwhile, fill a large pot with water and bring to a boil over
high heat, then add ¼ cup salt. Stir in the pasta. Cook, stirring
frequently to keep the pasta from sticking to the bottom of the pot,
for 3 minutes fewer than what the box says for al dente. Reserve
1 cup of the pasta cooking water, then drain the pasta and transfer
it to the pot with the sauce.

Increase the heat to medium-high, add the cheese, and cook,
tossing constantly and adding splashes of the pasta cooking water
as needed to ensure every noodle is coated in the loose, glossy,
not-sticky, thick sauce, about 3 minutes. Remove from the heat.
Add the reserved guanciale and toss to combine. Serve each bowl
of pasta with black pepper, a drizzle of good olive oil, and a heavy
sprinkle of cheese.

FLUFFY SHEET PAN FOCACCIA

Makes 1 roughly 16 × 12-inch focaccia

3 cups warm water
(105° to 115°F)

1 tablespoon honey

1 (¼-ounce) envelope active
dry yeast (2¼ teaspoons)

6 cups all-purpose flour or
bread flour

2 tablespoons kosher salt

½ cup extra-virgin olive oil,
plus more as needed

Nonstick cooking spray
(optional)

Optional additions: 2 cups
pitted olives; 1 onion or
shallot, thickly sliced; 8 to 12
ounces fresh tomatoes, cut
into bite-size pieces; a few
sprigs of woody herbs, such
as rosemary, oregano,
or marjoram

Flaky sea salt

MAKE-AHEAD: *Dough—up to
24 hours*

PAIR WITH: *Clam & Corn Pasta
(page 141), Amatriciana for a
Crowd (page 219), Party Chicken
with Feta & Fennel (page 27),
Veg & Dip Spread of Dreams
(page 38)*

Don't be scared: Making light, airy, olive oil–laden focaccia is so much easier than you think (I promise!). Serve this in a stunning aperitivo spread alongside olives, nuts, cheese, and bread, and/or with a main, for sopping up sauces. Focaccia is best eaten immediately, when warm—or at least the same day it's baked. But you can prep your dough and let it rise in the refrigerator the day before, so most of the prep is done in advance.

In a medium bowl, whisk together the warm water, honey, and yeast. Let sit for 5 to 10 minutes, until the surface is foamy.

In a large bowl, whisk together the flour and salt. Add the yeast mixture to the flour mixture and stir until a moist, lumpy dough comes together with no dry bits of flour remaining.

Pour ¼ cup of the olive oil into a separate large bowl. Rub a bit of oil on your hands and scrape off any dough stuck to the spoon, then scoop the dough into the bowl with the oil and turn to coat completely. Lightly oil a tight-fitting lid or plastic wrap and cover the bowl. If baking same day, let stand in a warm spot until the dough has more than doubled in size, 3 to 4 hours. (*Alternatively, if baking the next day, refrigerate for at least 8 hours or up to 24 hours to let the dough rise slowly.*)

Pour 3 tablespoons of the oil onto a rimmed baking sheet and use your fingers or a pastry brush to coat the bottom and sides of the pan. Uncover the dough and lightly oil your hands. Starting from one side of the bowl, scoop under the dough and gently fold it over itself, deflating it slightly. Rotate the bowl one quarter turn and repeat scooping and folding three times, until the dough is a slightly deflated mound. Turn out the dough (and any oil from the bowl) onto the prepared baking sheet, gently stretching it to roughly the same size as the pan—but don't force it to the edges (it will tear). Let the dough sit in a warm spot until doubled in size and very aerated, anywhere from 30 minutes to 3 hours, depending on the temperature of your kitchen.

Position a rack in the center of the oven and preheat to 450°F. Lightly oil your fingers and press them into the dough, all the way to the bottom of the pan, dimpling it all over. Scatter any additions evenly over the top of the dough, lightly pressing to adhere. Drizzle the remaining 1 tablespoon oil over the top and sprinkle with flaky salt. Bake the focaccia until deeply golden brown and puffed, 18 to 22 minutes, rotating halfway through. Let cool slightly before slicing. Serve immediately.

A NOODLE SOUP TO GET PEOPLE EXCITED

Serves 6

3 tablespoons neutral oil, plus more as needed

1 pound ground chicken, turkey, or pork, or 10 to 20 ounces firm or extra-firm tofu, crumbled

6 garlic cloves, grated

1 (2-inch) piece fresh ginger

Kosher salt

5 tablespoons soy sauce, plus more as needed

2 teaspoons fish sauce (optional)

1¼ cups tahini

4 cups low-sodium chicken stock or vegetable stock

3 tablespoons white or red miso paste

4 to 6 tablespoons chili crisp, to taste, plus more for serving

1 large bunch greens, such as kale, torn

8 ounces long noodles, such as soba, ramen, or rice vermicelli

Seasoned or unseasoned rice vinegar, for serving

Thinly sliced scallions, for serving

Toasted sesame seeds, for serving

PAIR WITH: *Gochugaru-Spiked Veg (page 170), Bulgogi-ish Lettuce Wraps (page 169)*

In the dead of winter, we all deserve food that makes us feel something. One of the dishes that does just that for me is the Mera Mera Soba at Cocoron, an always-busy restuarant on Delancey Street in downtown Manhattan. I daydream about this noodle soup and its bubbling, rich, red-orange broth—the ultimate cure for gloomy days. It's the dish that inspired this recipe: a soup to get friends excited, even in the most wintry weather, to come over for dinner.

Heat the oil in a large pot over medium-high heat. When the oil is shimmering, add the chicken and use a wooden spoon to spread it into an even layer across the bottom of the pot. Cook, undisturbed, until deeply golden brown on the bottom, 4 to 6 minutes. Add the garlic, ginger, and a big pinch of salt. Reduce the heat to medium. Use the spoon to break up the meat and stir until the chicken is just cooked through, 2 to 4 minutes. Use a slotted spoon to transfer the mixture to a medium bowl, leaving as much fat in the pot as possible (if there's none, add a small splash of oil).

Add the soy sauce and fish sauce (if using) to the reserved fat in the pot and let sizzle and caramelize over medium-high heat until slightly thickened, 1 to 2 minutes. Reduce the heat to medium-low. Add the tahini and slowly whisk in 1 cup water, ensuring the mixture stays smooth. Add another 2 cups of water and the stock. Continue to cook until the mixture is warmed through (but not simmering), about 15 minutes, reducing the heat as needed.

In a small bowl, whisk together the miso and chili crisp. Stir in ¼ cup of the warmed broth, then add the mixture to the pot. Add the kale to the pot. Cook, stirring occasionally, until the kale has softened and the flavors have melded, 20 minutes more (don't let the soup come to a simmer, or it will separate).

Meanwhile, cook the noodles according to the package directions. Drain and rinse with cold water to stop the cooking and prevent them from sticking together.

Stir the chicken mixture into the soup and cook until the chicken is warmed through, about 7 minutes. Taste the broth and add more soy sauce as needed. Keep warm over low heat.

Divide the noodles among large or deep bowls, then ladle the soup over the noodles. Add a splash of vinegar for a little acidity. Serve topped with scallions, sesame seeds, and more chili crisp.

CREAMY TOMATO SOUP

Serves 6 to 8

When in doubt, tomato soup (with nonnegotiable grilled cheese) is my answer to midwinter, midweek "blahs." Does this whole meal give a little bit of an after-school-special vibe? Absolutely. But it's the comforting hit of nostalgia I crave every now and then (okay, more like all the time).

2 tablespoons extra-virgin olive oil, plus more for serving

1 medium yellow or red onion, diced

6 garlic cloves, diced, plus 1 for serving, grated

Kosher salt and freshly ground black pepper

½ teaspoon red pepper flakes (optional)

2 (28-ounce) cans crushed tomatoes

1 (12-ounce) jar roasted red peppers, drained and roughly chopped

½ teaspoon sugar

¾ cup heavy cream or well-shaken canned full-fat coconut milk

½ bunch fresh dill and/or parsley, finely chopped, for serving

Heat the oil in a large pot over medium heat. When the oil is shimmering, add the onion and garlic. Season with a big pinch of salt, black pepper, and red pepper flakes (if using).

Cook, stirring often, until the onion is soft and translucent, 5 to 8 minutes. Stir in the crushed tomatoes, roasted red peppers, 2 cups water, the sugar, and 2 teaspoons salt. Bring the mixture to a simmer over medium heat, cover, and cook until the flavors have melded, about 15 minutes. If you like smooth soup, transfer to a heatproof blender and, working in batches as needed, puree until silky, then return to the pot. (Alternatively, use an immersion blender.) Stir in the cream and cook, uncovered, over low heat until warmed through, 1 to 2 minutes. Taste and add more salt and pepper as needed.

In a small bowl, combine the grated garlic and the dill and season with a big pinch of salt. Rub the mixture together with your fingertips to create a fragrant soup topper. Serve the soup immediately with more black pepper on top, plus a handful of the herb mixture and an extra drizzle of oil.

FOCACCIA GRILLED CHEESE

Serves 8 to 10

1 pound sharp yellow or white
    cheddar or pepper
    Jack cheese

12 to 16 ounces Gruyère cheese

1 focaccia (around 16 × 12
    inches), homemade
    (page 220) or store-bought

Freshly ground black pepper

4 tablespoons (½ stick)
    unsalted butter, at
    room temperature

Grilled cheese is a perfect meal for one. But I wasn't going to let that stop me from developing a grilled cheese recipe that can (easily) feed a group. Instead of making individual sandwiches on the stovetop, using your own homemade focaccia is not only extremely impressive, it's also a strategic choice that gets you one big sandwich that's ready for everybody at the same time. And if you're up for a little multitasking, it even cooks in the oven while your tomato soup bubbles on the stove.

Position a rack in the center of the oven and preheat to 400°F. Line a baking sheet with parchment paper.

Grate the cheese on the large holes of a box grater or in a food processor fitted with the coarse grating disk.

Place the focaccia on a large cutting board. Using a serrated knife, trim (and eat) the edges of the focaccia to make them straight and easier to slice. Carefully split the focaccia in half horizontally, keeping your knife level with the cutting board to get two similarly thick halves.

Place the bottom half on the prepared baking sheet. Evenly sprinkle the grated cheese all over the bottom half of the focaccia, then season the cheese with pepper. Place the focaccia top over the bottom half, cut-side down, then spread the top of the sandwich with butter.

Place a second piece of parchment on top of the sandwich, then set a second baking sheet, flat-side down, on the parchment. Place two heavy oven-safe skillets (cast iron is great for this, but use whatever will fit) on top of the baking sheet to weigh down the sandwich, then transfer this whole rig to the oven. Bake until the cheese is melted and the sandwich is pressed, 20 to 25 minutes. Carefully remove the skillets and baking sheet and pull off the top piece of parchment. If you'd like to add a bit more color and crunch to the top, switch the oven to broil and return the sandwich to the oven for 2 to 4 minutes, until the top is deeply golden.

Let the sandwich cool in the pan for 5 minutes, then carefully slide it out onto the biggest cutting board you have. Use a serrated knife to cut the sandwich into 8 or 16 pieces.

A STEAKHOUSE DATE FOR TWO

I'VE LIVED IN NEW YORK CITY FOR OVER A DECADE, but I've never once been to a New York City restaurant on Valentine's Day. New York City restaurants are some of the most romantic places in the world. And yet.

When it comes to date night—whether that's on Valentine's Day or any other day of the year—I prefer going big at home. Dial up the little luxuries that are especially worth it when it's just two of you. Have a Scallop Snack, *while* you cook. Get the thick-cut bacon to top your wedge with. Spring for the best (and, more important, your favorite) cut of meat at the butcher. Find a bottle of wine you can't wait to share. Make a chocolate mousse and eat it right out of the bowl, why don't you.

Not only is this meal ideal for sharing—it's also ideal for cooking with someone. One of you should start by seasoning the meat while the other gets to scrubbing and slicing potatoes. Let both sit for at least half an hour—this would be the time to enjoy your Scallop Snack—and then pop the potatoes into the oven. About halfway through their baking time, heat up your pan and get to cooking your steak. When you pull your potatoes from the oven, immediately season them. Then, while the steak rests, dress your wedge salad. *Et voilà*, suddenly you have a rib eye, fries, and a wedge salad—your own personal steakhouse date for two.

Menu

For 2

**MY PERFECT MARTINI
(AND YOURS)**

**A SCALLOP SNACK
WHILE YOU COOK**

VALENTINE WEDGE

STEAK FRITES FOR TWO

**BOOZY CHOCOLATE MOUSSE,
TO SHARE**

A SCALLOP SNACK WHILE YOU COOK

Serves 2

1 tablespoon extra-virgin
    olive oil

3 tablespoons unsalted butter,
    cut into pieces

½ pound sea scallops (ideally
    dry-pack), side
    muscle removed

Kosher salt

2 tablespoons dry vermouth,
    white wine, or fresh
    lemon juice

Pinch of red pepper flakes
    (optional)

File this one under "things you can't do when you're hosting a lot of people." Good scallops need so little in the way of cooking prep, this recipe is basically all reward after only a few minutes of work. Just promise me you'll take a pause from cooking to enjoy them. I'd tell you to sit down, but standing at your kitchen counter is more than acceptable.

In a large cast-iron or stainless-steel skillet, heat the olive oil and 2 tablespoons of the butter over medium-high heat. Pat the scallops dry with paper towels and season both sides with salt. When the butter has melted and is beginning to foam, add the scallops and cook, undisturbed, until a golden crust forms on the bottom, 2 to 3 minutes.

Flip the scallops. Add the remaining butter, the vermouth, and the red pepper flakes (if using) to the pan. Let the butter sizzle for 30 seconds, then carefully nudge the scallops toward one edge of the pan, tilting the pan so the buttery mixture pools opposite them. Use a spoon to scoop up the melted butter and continuously pour it over the scallops for about 1 minute.

Transfer the scallops to a shallow bowl, then pour the pan sauce over the top. Enjoy immediately.

VALENTINE WEDGE

Serves 2, easily scaled up

FOR THE DRESSING

½ cup crumbled blue cheese, plus more for serving

Zest and juice of 1 lemon

1 garlic clove, grated

½ small bunch dill, chopped, plus more for serving

¼ cup mayonnaise

¼ cup sour cream

3 tablespoons buttermilk

Kosher salt and freshly ground black pepper

FOR THE SALAD

2 slices thick-cut bacon, cut into 1-inch pieces

8 ounces cherry or grape tomatoes, halved, or beefsteak tomatoes, thinly sliced

Kosher salt

½ large head iceberg lettuce, wilted outer leaves discarded

Lemon wedges, for serving

Chopped fresh chives, for serving

Flaky sea salt, for serving

MAKE-AHEAD: *Dressing—up to 1 day*

PAIR WITH: *Meatballs & the Reddest Sauce (page 251), Big Calzone Night (page 157)*

It's not a steakhouse dinner if it doesn't start with a wedge salad. And this is exactly the wedge salad I wish every steakhouse served. Be sure to keep your lettuce in the fridge until the moment you're ready to serve—you want to keep it as cold as possible for the best wedge experience. The crispness of the leaves, the creaminess of the dressing, the smoky porky-ness of the bacon, and the pop of cherry tomatoes make for a combination I probably think about a little too much.

MAKE THE DRESSING: In a medium bowl, combine the blue cheese, lemon zest, lemon juice, garlic, and dill. Add the mayonnaise, sour cream, and buttermilk and season with a big pinch of salt and a few grinds of pepper. Stir to combine well. Taste and add more salt and pepper as needed. *Refrigerate until you're ready to serve, up to 1 day.*

MAKE THE SALAD: Line a plate with paper towels. In a cold nonstick or cast-iron skillet, arrange the bacon pieces in a single layer. Cook over medium heat, flipping often, until the bacon is browned and crisp, 7 to 9 minutes. Remove the skillet from the heat and use a slotted spoon to transfer the bacon to the prepared plate to drain. Pour any bacon fat left in the skillet into a heatproof container and reserve.

In a small bowl, toss the tomatoes with a big pinch of kosher salt and 2 teaspoons of the rendered bacon fat. (Store the remaining bacon fat in its airtight container in the refrigerator for another use; it will keep for up to 3 months.)

Halve the iceberg through the core so you have two large wedges, then cut each wedge in half crosswise. Swipe a big spoonful of dressing onto a serving plate, then place the iceberg wedges on top of the dressing. Squeeze 1 of the lemon wedges over the lettuce, season with salt, and top with a few more large dollops of dressing. Scatter the tomatoes and bacon (crumbling it further, if you like) over the dressing. Top with more crumbled blue cheese, more dill, chives, and flaky salt. Serve with the remaining dressing and lemon wedges on the side.

STEAK FRITES FOR TWO

Serves 2

Perhaps the highest calling of meat and potatoes, steak frites is the sexy French version that's also a perfect date night dish to share. Rather than a typical thin, bistro-style fry, we're going for a thicker cut here, because I don't personally want to do any deep-frying on date night (if that's your style, I love that for you and would also recommend you up the ante and go for an onion ring while you're at it). These fries cook in the oven in an oil bath and come out crispy on the outside, fluffy on the inside, just waiting to be paired with an expertly seared steak. I'm talking about you—you're the expert. I have faith in you.

FOR THE STEAK AND FRIES

1 (1½- to 2-pound) bone-in rib eye steak, 1 to 1½ inches thick

Kosher salt and freshly ground black pepper

1 pound large Yukon Gold potatoes (about 4), scrubbed

4 tablespoons neutral oil

2 tablespoons unsalted butter

FOR THE DIPPING SAUCE

1 garlic clove, grated

½ cup crème fraîche or sour cream

2 tablespoons whole-grain mustard

2 teaspoons fresh lemon juice, plus more as needed

A few dashes of Worcestershire sauce (optional)

Kosher salt and freshly ground black pepper

MAKE-AHEAD: *Seasoned steak— up to 48 hours*

MAKE THE STEAK AND FRIES: On a plate, small baking sheet, or cutting board, season the steak all over with salt (about 1 teaspoon per pound of meat) and pepper. Let sit at room temperature for at least 30 minutes or up to 2 hours. (*Alternatively, season the steak in advance and store in the refrigerator, uncovered, for up to 48 hours; let sit at room temperature for at least 30 minutes before searing.*)

Meanwhile, halve each potato lengthwise, then slice each half into ½-inch-thick wedges. Place in a large bowl of hot water and let soak for 30 minutes.

Position a rack in the center of the oven and preheat to 425°F.

Drain the potatoes and dry well. Place on a baking sheet, toss with 3 tablespoons of the oil and ½ teaspoon salt, then arrange in a single layer, ensuring they're not touching. Bake, flipping halfway through, until the potatoes are deeply golden on all sides, 30 to 35 minutes. Remove from the oven and immediately season with another pinch of salt and a couple of grinds of pepper, too, if you'd like.

Meanwhile, in a medium cast-iron or stainless-steel skillet, heat the remaining 1 tablespoon oil over high heat. Pat the steak dry with paper towels. When the oil begins to smoke and ripple, carefully add the steak to the pan. Cook, undisturbed, until deeply browned on the bottom, 4 to 6 minutes (go for the lower end if you like steak on the medium-rare side). Use tongs to flip the steak (it should release from the pan immediately; if it doesn't, cook for 1 to 2 minutes more until it does) and reduce the heat to medium-high. Cook until the other side is deeply browned, about 5 minutes. Use tongs to lift the steak onto its fat cap—that's the longer side that's covered in pinkish-white fat—and cook for 1 minute to melt the fat. Return the steak to the second side, add the butter to the skillet, and turn off the heat. Carefully nudge the steak toward one edge

of the pan, then tilt the pan so the butter pools opposite the steak. Use a spoon to scoop up the melted butter and continuously pour it over the steak for about 1 minute. Transfer the steak to a cutting board and let it rest for at least 10 minutes or up to 1 hour before slicing.

MEANWHILE, MAKE THE DIPPING SAUCE: In a small bowl, combine the garlic, crème fraîche, mustard,

lemon juice, and Worcestershire (if using) and season with a big pinch of salt and a few good grinds of pepper. Stir together, then taste and add more salt or lemon juice as needed.

Slice the steak against the grain into ½-inch-thick pieces. Divide the sliced steak and fries between two plates (or share a plate, if you prefer, you romantics), and serve with the dipping sauce alongside.

BOOZY CHOCOLATE MOUSSE, TO SHARE

Serves 2

4 ounces bittersweet
    chocolate (70% cacao),
    roughly chopped

Kosher salt

4 teaspoons crème de cassis,
    medium amaro (such as
    Averna or Montenegro), or
    Grand Marnier (optional;
    see Note)

½ cup heavy cream or sour
    cream, for serving (optional)

Flaky sea salt, for serving

NOTE: If you don't have any of these liqueurs or you'd prefer not to use alcohol, omit it and use 4 teaspoons water instead.

MAKE-AHEAD: *Up to 1 day*

After cooking a multi-dish dinner, the last thing I want to do is to spend another hour making dessert. My capacity for an in-the-moment dessert-making activity maxes out at around 10 minutes and "extremely easy," which is exactly what this chocolate mousse is. It comes together so quickly, and requires so few ingredients, you could decide on an impulse to make it—instant Big Night achieved.

Fill a large bowl with ice and a little cold water. In a medium saucepan, combine the chocolate, ⅓ cup water, a pinch of kosher salt, and the crème de cassis (if using). Cook over medium heat, stirring, until the chocolate has melted and the mixture is smooth, 2 to 3 minutes.

Immediately pour the melted chocolate mixture into a medium bowl and place it over the ice bath. Vigorously whisk (or beat with a hand mixer on medium speed) until the mixture is thick, 2 to 4 minutes. It will seem like nothing is happening at first, but then the mixture will firm up very quickly; if it becomes too firm for your liking, warm the mousse over a medium pot of simmering water to slightly melt, then rewhip over a fresh ice bath.

Divide the mousse between two bowls or cups, if you like, or share it straight from the bowl. (*If not serving immediately, transfer to your desired serving vessel, cover, and refrigerate for up to 1 day.*)

If you want to top the mousse with something creamy, rinse the mixer beaters, then pour the cream into a medium bowl. Beat on medium speed until the cream holds medium peaks, about 1 minute (if using Greek yogurt or sour cream, simply dollop it on the mousse). Top the mousse with the whipped cream, or swirl or fold it into the mousse. Finish with a pinch of flaky salt and serve in individual glasses—or eat it right out of the bowl.

MY PERFECT MARTINI

Makes 1 drink

2¾ ounces gin
  (I like Tanqueray)

1 ounce dry vermouth
  (I like Dolin)

½ ounce Castelvetrano
  olive brine

Pitted Castelvetrano olives,
  for serving

Martinis are not a dinner party drink. Just because you can batch them, does not mean you should. I believe a martini should *always* be made to order—whether I'm in a restaurant or hosting in my own home. The magic of a martini is that, like your morning coffee or your astrology chart, it's personal: your own individualized experience.

So while I do not recommend making martinis en masse, I do believe that one of the best ways to make someone feel loved and seen is to make them their own martini. *Exactly* how they like it. If you're reading this thinking, *But what if I don't know how I like it?* let's change that. You'll only find out by trying. Start here, if you'd like. Then turn the page to learn how to modify this recipe until you find your own personal, perfect martini. My Perfect Martini: *gin, slightly dirty, shaken hard, with olives.*

Place your glass in the freezer the minute you think, *I might like a martini.*

Fill a cocktail shaker halfway with ice. Pour in the gin, vermouth, and olive brine. Cover the shaker and shake until very, very cold, about 20 seconds (some people will say this is incorrect, but I prefer to have the tiniest shards of ice in my drink) or until your hand starts hurting because it's so cold.

Grab your chilled glass out of the freezer. Pour the martini into the glass and garnish with one Castelvetrano olive (or three, on a cocktail pick). Drink it before it warms up.

five questions to help you find your perfect martini

GIN OR VODKA? . . . .

This is the primary ingredient of this cocktail, and, as such, it's the choice that's most important. Gin will give you a more botanical flavor profile, while vodka tastes cleaner. If you have a favorite brand, stick with it.

DRY OR WET? . . . .

Vermouth, a fortified wine, imparts its own flavor, separate from the gin or vodka. It also has a slightly lower ABV than either base spirit. "Dry" means less vermouth; "wet" means more. The drier your martini, the more spirit is poured in its place, resulting in a boozier martini. Try your 'tini "50-50" to equalize the vermouth and spirit volumes and create a slightly less boozy cocktail.

DIRTY (OR NOT)? . . . .

Cocktail purists will tell you that a dirty martini is truly a different cocktail than a martini, and I get why, because a lot of people translate "dirty" as "dump a lot of olive juice into that drink." I like my martinis only very slightly dirty—so there's a whisper of briny flavor, but it's not at all overpowering.

SHAKEN OR STIRRED? . . . .

Stirred is classic. I like mine shaken, because it guarantees my martini will end up very, very, very cold—and I also like the little ice particles that you get in your drink when you shake it. A bonus (for me!) is that shaking dilutes the martini more than stirring.

WHAT GARNISH? . . . .

Choose your fighter: lemon twist, olives, or onion. Mine will always be olives, but I think a lemon peel is lovely and an onion is bold.

RED SAUCE FOR THE HOLIDAYS, RED SAUCE FOR EVERYONE

AROUND THE HOLIDAYS, comfort—and comfort food—looks a little different for everyone. When I'm holiday hosting, I find myself craving the familiar (but still festive). I look for low-stress dishes. But I also want crowd-pleasers, because the greatest gift of all is . . . compliments. Just kidding—the greatest gift is for people to be completely blown away by how good the food tastes and how good it makes them feel. This Bigger Night is a nod to the red-sauce Italian restaurants that make me feel like I'm already at a party from the moment I sit down in the big red leather booth.

It's the holidays, so we're going Big menu-wise, but don't worry, we have a plan. Our two main dishes—Meatballs & the Reddest Sauce and Side Lasagna (yes, that's right, it's a SIDE of lasagna)—share the same sauce, so once you've made that, which you can absolutely do in advance, you're well on your way. The focaccia can be prepped ahead of time before it's turned into focaccia garlic bread, and same goes for the crème brûlée. You can even blanch your greens and make your salad dressing early in the day, so all that's left to do is toss. It's the holidays—the last thing I want is for you to have to hide out in your kitchen all night. Unless, that is, you want to.

Menu

For 8

BITTER GREENS & BROCCOLI CAESAR

FOCACCIA GARLIC BREAD

MEATBALLS & THE REDDEST SAUCE

SIDE LASAGNA

FAMILY-STYLE CRÈME BRÛLÉE

BATCHED HANKY-PANKYS

BITTER GREENS & BROCCOLI CAESAR

Serves 6 to 8 as an appetizer or side, or 3 or 4 as a main

Kosher salt

2 to 2½ pounds broccoli rabe or broccolini, or a mix (2 to 3 bunches)

⅔ cup plus 3 tablespoons extra-virgin olive oil

1 cup panko breadcrumbs

2 to 3 large lemons

Freshly ground black pepper

5 oil-packed anchovy fillets

3 garlic cloves, peeled

4 large egg yolks (see Note)

2 teaspoons Worcestershire sauce

3 tablespoons Dijon mustard

½ cup freshly grated Parmesan cheese, plus more for serving

2 heads radicchio or Castelfranco, roughly chopped

NOTE: If you can't eat raw egg or you'd simply prefer not to, swap the yolks for 3 tablespoons mayonnaise.

MAKE-AHEAD: *Blanched broccoli rabe, breadcrumbs, and dressing—up to 2 days*

PAIR WITH: *Crispiest Chicken Milanese (page 150), Big Calzone Night (page 157), Amatriciana for a Crowd (page 219)*

I will take a Caesar salad any which way it is served to me. But this is my ultimate, desert-island Caesar salad. Bitter, bracing broccoli rabe and radicchio meets punchy, rich yet lemony dressing, with a shower of perfectly browned breadcrumbs to complete the trifecta of textural heaven. From a dinner party perspective (yes, there would be dinner parties on my desert island), it's even better, because every single component can be prepared in advance. All you have to do when you're ready to serve is combine and toss.

Fill a large pot with water and bring to a boil over high heat, then add a few big pinches of salt. Fill a large bowl with water and ice. Working in batches as needed, add the broccoli rabe to the boiling water and cook until just tender, about 30 seconds. Using tongs or a spider, transfer the broccoli rabe to the ice bath. Let cool completely, 5 to 10 minutes. Drain and dry well, then roughly chop. *Store in an airtight container in the refrigerator for up to 2 days; let come to room temperature before serving.*

Heat 3 tablespoons of the olive oil in a large skillet over medium heat. When the oil is shimmering, add the panko and cook, stirring often, until deeply golden brown, about 4 minutes. Remove from the heat. Zest 1 lemon into the breadcrumbs, then add a big pinch of salt and lots of pepper. Stir to combine. Taste and season with salt. Transfer the breadcrumbs to a paper towel–lined plate and set aside. *Store in an airtight container in the refrigerator for up to 2 days; let come to room temperature before serving.*

Finely chop the anchovies, then sprinkle with salt and use the side of your knife to mash them into a paste.

Zest the remaining lemon into a separate medium bowl. Grate the garlic into the bowl as well. Juice both lemons until you have ½ cup and whisk the lemon juice, egg yolks, Worcestershire, and mustard into the bowl. Whisking constantly, very slowly add the remaining ⅔ cup olive oil. Continue to whisk until the mixture is emulsified and looks like thick heavy cream, about a minute, then whisk in the anchovy paste, Parmesan, 1 teaspoon salt, and lots of pepper. Taste and add more salt and pepper as needed—it should be very well seasoned and strongly flavored. *Store the dressing in an airtight container in the refrigerator for up to 2 days.*

To serve, in a large bowl, toss the radicchio and broccoli rabe with the dressing, then transfer to one or two large serving plates. Top with the breadcrumbs, more Parmesan, a drizzle of olive oil, and some pepper and serve immediately.

FOCACCIA GARLIC BREAD

Serves 6 to 8

10 garlic cloves, smashed and peeled

4 oil-packed anchovy fillets (optional)

Kosher salt

½ cup (1 stick) unsalted butter, at room temperature (it should be spreadable)

2 tablespoons dried oregano or parsley

2 tablespoons extra-virgin olive oil

½ teaspoon freshly ground black pepper

¼ teaspoon red pepper flakes (optional)

1 focaccia (about 16 × 12 inches), homemade (page 220) or store-bought

MAKE-AHEAD: *Focaccia—up to 24 hours*

PAIR WITH: *Amatriciana for a Crowd (page 219), Clam & Corn Pasta (page 141), Creamy Tomato Soup (page 224)*

Garlic bread reaches its peak form when it's made with focaccia. That said, whether you're using focaccia you made yourself, focaccia you purchased in a store, or even another kind of bread altogether (a baguette or other crusty loaf also works great), this should mostly serve as your reminder that garlic bread is an attainable luxury any night of the week.

Position a rack in the center of the oven and preheat to 350°F. Line a baking sheet with parchment paper.

On a cutting board, sprinkle the garlic and anchovies (if using) with a pinch of salt and use a knife to smash and chop them until the mixture is very finely chopped, almost to a paste. Transfer the garlic mixture to a small bowl and use a fork to smash and stir in the butter, oregano, olive oil, black pepper, red pepper flakes (if using), and another big pinch of salt until combined and roughly the consistency of buttercream frosting. (Alternatively, use a food processor or a mortar and pestle for this whole process.)

Place the focaccia on a large cutting board. Using a serrated knife, trim (and eat) the edges of the focaccia to make them straight and easier to slice. Carefully split the focaccia in half horizontally, keeping your knife level with the cutting board to get two similarly thick halves.

Place the focaccia halves cut-side up on the prepared baking sheet. Use a spatula to spread about two-thirds of the garlic butter over the cut side of the bottom half. Place the top half over, cut-side down, then spread the remaining garlic butter over the top. Tightly wrap the baking sheet with foil.

Bake until very fragrant and warmed through, 15 to 20 minutes. Remove the foil and increase the oven temperature to 425°F. Bake until the focaccia is toasty on top, about 5 minutes more.

Let cool for 5 minutes before slicing into as many pieces as you'd like. Serve warm.

MEATBALLS & THE REDDEST SAUCE

Serves 6 to 8 (makes about 20 meatballs)

2 tablespoons extra-virgin olive oil

¾ cup panko breadcrumbs

2 tablespoons dried oregano

2½ teaspoons kosher salt

2 teaspoons freshly ground black pepper

2 large eggs

1 cup whole-milk ricotta

1 small yellow onion, minced

8 garlic cloves, grated

1 pound ground beef (no more than 85% lean)

1 pound ground pork

8 cups The Reddest Sauce (recipe follows)

Freshly grated Parmesan cheese, for serving

NOTE: If tonight's not a Side Lasagna night, your meatballs are waiting for spaghetti: Cook 1 pound of spaghetti (or other long noodles) to just under al dente. Reserve 1 cup of the pasta water. Drain the noodles, return them to the pot, and finish cooking the pasta over low heat, adding a bit of Reddest Sauce, pasta water, and Parmesan, until everything is saucy and glossy.

I know I'm not the first person to say this, but that doesn't make it any less true: The best tomato sauces are made with the best tomatoes. My favorite canned tomatoes are Bianco DiNapoli's, from California, and imported San Marzanos are excellent, too. And when I say *Reddest*, I mean it. These meatballs need about 8 cups of Reddest Sauce—use the additional 4 cups for A Side Lasagna (page 255) or spaghetti (see Note), or freeze for a future Red Sauce Night.

Position a rack in the center of the oven and preheat to 400°F. Drizzle a baking sheet with olive oil.

In a small bowl, stir together the panko, oregano, salt, and pepper.

Crack the eggs into a large bowl and lightly beat. Add the ricotta, onion, and garlic. Stir to combine. Add the panko mixture, along with the ground beef and pork. Use your hands to gently combine the mixture completely—take care to ensure everything is evenly distributed and no streaks of egg remain, but don't overknead. Using your hands, pull off pieces of the mixture and very gently shape into 2-inch meatballs, arranging on the prepared baking sheet (you should have about 20).

Bake the meatballs until nearly cooked through (at least 150°F on an instant-read thermometer plunged into one), 10 to 12 minutes. Turn your broiler on and broil the meatballs for 1 to 2 minutes, until they've taken on a bit of color.

Meanwhile, warm the sauce in a medium pot over medium-low heat, if necessary. Taste the sauce and add more salt and pepper as needed. Cover and keep warm over low heat.

Plop the finished meatballs into the warm sauce and increase the heat to medium-low. Cook, uncovered, until the meatballs are cooked through (at least 165°F), about 10 minutes. Transfer the meatballs and as much sauce as you'd like (there will be extra!) to a shallow serving bowl. Top with Parmesan and serve warm.

→

The Reddest Sauce

Makes about 12 cups

½ cup extra-virgin olive oil

12 garlic cloves, diced

2 medium yellow onions, diced

Kosher salt and freshly ground black pepper

1 (6-ounce) can or (4½-ounce) tube tomato paste, preferably double-concentrated

½ teaspoon red pepper flakes (optional)

3 (28-ounce) cans crushed tomatoes or whole peeled tomatoes, crushed by hand (see Note)

2 teaspoons sugar

NOTE: There are several ways to go about the tomatoes in this recipe. The simplest would be to use crushed tomatoes straight out of the can. I am a control-freak about my tomatoes, and as such, I like to use whole and crush them myself. You can do so in a large bowl, with your hands or an immersion blender, or in the can itself, snipping them with kitchen shears until you've reached your ideal consistency. For this sauce, I'm aiming for tomato pieces no bigger than ½ inch. A fully smooth sauce would also be perfectly delicious, just a different textural experience.

MAKE-AHEAD: *Up to 1 week (or freeze up to 6 months)*

Heat the oil in a medium pot over medium heat. When the oil is shimmering, add the garlic and onions and season with a big pinch of salt and lots of black pepper. Cook, stirring often, until the onions are translucent but not browned, about 5 minutes. Add the tomato paste and red pepper flakes (if using) and cook, stirring often, until the tomato paste has darkened to a brick-red color, about 2 minutes. Add the tomatoes, sugar, 2 teaspoons salt, and 1 cup water. Bring the mixture to a boil, then reduce the heat to medium-low, partially cover the pot, and simmer, stirring occasionally, until the sauce has thickened and tastes amazing, about 1 hour. *The sauce can be cooled and stored in airtight containers in the refrigerator for up to 1 week or frozen for up to 6 months. Defrost and warm before serving.*

SIDE LASAGNA

Serves 6 to 8 as a side, or 3 or 4
as a main

Extra-virgin olive oil

Kosher salt

8 ounces lasagna noodles
　(not no-cook)

1 cup whole-milk ricotta

2 cups shredded fresh or
　low-moisture mozzarella,
　plus more for sprinkling

1 cup freshly grated Parmesan
　cheese, plus more
　for sprinkling

Freshly ground black pepper

4 cups The Reddest Sauce (see
　page 252) or high-quality
　store-bought marinara
　sauce (I like Ciao Pappy
　and Rubirosa)

NOTE: You can bake this in
any 1½-quart vessel—glass,
ceramic, even a cake or
loaf pan. Depending on
that vessel, you might have
sauce, ricotta, and/or pasta
left over—this makes for the
perfect private bite with just
you, your noodles dipped
in sauce, and cheese. Or
store it in airtight containers
in the refrigerator for a
deconstructed lasagna
snack the next day.

MAKE-AHEAD: *Up to 48 hours*

PAIR WITH: *Crispiest Chicken
Milanese with Spicy Balsamic
Arugula (page 150)*

I love lasagna. Many people love lasagna. But lasagna is high-
pressure! There's the sauce—a project in and of itself—the fillings, the
layers and layers, and the prized gooey-yet-slightly crunchy top. It's
a labor of love, and at the end of the process, I always find myself
wanting . . . more? Not more lasagna, per se, but something else on
my plate, too. After all that effort, you get one singular (very delicious,
but singular) dish. When I cook for people, I want to give them many
flavors and textures. And that is why I present to you: the Side of
Lasagna. Even just saying it sounds like a mic drop. You didn't make
lasagna—you made it as a SIDE.

Position a rack in the center of the oven and preheat to 400°F. Drizzle
a baking sheet with oil.

Bring a large pot of water to a boil over high heat. Add 2 tablespoons
salt and the lasagna noodles. Cook, stirring often, until barely al
dente, about 4 minutes. Reserve ⅓ cup of the pasta cooking water,
then drain the noodles and lay them on the oiled baking sheet to cool
(if you need to overlap them, drizzle with more oil to prevent sticking).

Meanwhile, in a medium bowl, stir together the ricotta, mozzarella,
Parmesan, reserved pasta cooking water, ½ teaspoon salt, and lots of
pepper. You will have a thick, delicious mixture.

Spread a large spoonful or two of the red sauce into the bottom of
a 1½ quart vessel, then add a layer of noodles, slightly overlapping
and/or trimming as needed. Spoon over ½ to ¾ cup of the sauce
(stick to the lower end if using a loaf pan, and the higher end if using
a wider dish), spreading it evenly. Dollop ½ to ¾ cup of the ricotta
mixture (same drill as the sauce) in small mounds all over, then add
another single layer of noodles. Repeat, ending with a final layer of
sauce and filling the vessel just about to the top but not stuffing it.
Sprinkle the top of the lasagna with more mozzarella and Parmesan.
Cover with foil and place on a clean baking sheet to catch drips.

Bake until bubbling, about 30 minutes, then uncover and bake until
very bubbly and the top layer of cheese and corners are beginning to
char, 20 to 30 minutes more. Let cool for at least 20 minutes or up to
2 hours before slicing in. Be patient here, as resting lets the juices settle
a bit and the flavors come together. *In fact, lasagna can be even
better the next day—let cool completely, then cover and refrigerate
for up to 48 hours. Reheat in a 350°F oven until warmed through and
gooey, about 20 minutes.*

FAMILY-STYLE CRÈME BRÛLÉE

Serves 6 to 8

3 cups heavy cream

1 teaspoon kosher salt

1 vanilla bean, scraped (see Note)

7 large egg yolks

¾ cup plus 1 to 2 tablespoons sugar

Boiling water

NOTE: Vanilla bean pods are pricey and often an unnecessary expense. But in a dessert like this, where vanilla is the main flavor, it's worth the splurge. As an alternative, you can use 1 tablespoon vanilla bean paste or pure extract; add it once you've removed the cream from the heat.

MAKE-AHEAD:

Custard—up to 48 hours

Cooked crème brûlée—up to 2 hours

Crème brûlée is one of those desserts that always looks pretty much exactly the same: little white ramekin on a crisp white tablecloth. To this, I say: *Free the brûlée!* You can make crème brûlée at home (no blowtorch required—your broiler will do the trick), and you don't need a bunch of little ramekins to serve it. I like to bring it straight to the table in the baking dish it cooks in and let everyone dig in and hack away at that caramelized-sugar top, lost in a shared dessert world together.

Position a rack in the center of the oven and preheat to 325°F. Place a 1½-quart glass baking dish inside another large baking dish or skillet (any shape/material will do, as long as the smaller one fits in the larger one with some room). Fill a teakettle or medium pot with water and bring to a boil.

In a medium saucepan, combine the cream and the salt. Use a paring knife to split the vanilla bean and scrape out the paste. Drop the paste and the pod into the saucepan. Turn the heat to medium-low and cook until hot and barely steaming but not yet simmering, 3 to 5 minutes—if it starts to bubble at the edges, reduce the heat. Remove the pan from the heat and discard the vanilla bean pod.

Meanwhile, in a large bowl, beat the egg yolks. Gently whisk in ¾ cup of the sugar until combined but not foamy. While whisking continuously (but gently), slowly pour about ½ cup of the hot cream into the egg mixture until completely combined. Use a rubber spatula to gently stir in the rest of the cream, about 1 cup at a time, stirring in a figure-eight motion until just combined. Strain through a fine-mesh sieve into a medium bowl or large liquid measuring cup. Pour the cream mixture into the small baking dish, then pour boiling water into the large baking dish until it comes three-quarters of the way up the sides of the small baking dish.

Bake until the custard is matte on top, set at the edges, and wiggles just a bit in the center, 35 to 50 minutes (if your small baking dish is deeper than 2 inches, the custard may take longer to set). Remove both dishes from the oven and let cool to room temperature, still nestled, about 1 hour. Remove the small baking dish from the water bath, dry the bottom, and cover the top with plastic wrap, taking care not to let the plastic hit the surface of the custard. Chill in the refrigerator until completely set, at least 3 hours or up to 48 hours.

When ready to finish the crème brûlée, remove the plastic wrap from the baking dish and let the custard sit at room temperature for 30 minutes. Heat the broiler.

Sprinkle the surface of the custard evenly with the remaining sugar, beginning with 1 tablespoon and adding more if the surface is not yet completely covered. Gently tilt and tap the outside of the dish to ensure the custard is evenly coated by a very thin layer of sugar—don't be tempted to add more. Return the baking dish to the oven and, keeping a close eye, broil until the top is deeply caramelized, 1 to 2 minutes. Serve immediately, or within 2 hours.

the amaro primer

I LOVE EVERYTHING ABOUT amaro (plural = amari): the beautiful bottles, the vast range of flavors, and the way it feels like the perfect cap to pretty much any meal. Most of all, I love what amaro does for a dinner party. Bring out a couple of bottles with or after dessert, and let people taste and choose their favorite. It's a conversation starter, it's a show-and-tell, it's a fun experience that your guests may not have at home, and it's a great way to stretch out the evening.

Amaro is a huge category of spirits. On one end of the spectrum, we have bottles like Aperol, which is most often associated with sweeter spritzes. On the other end of the spectrum, we have bottles like Fernet-Branca, which is tongue-curlingly bitter. And then we have everything in between—which is where the fun really lies, because there is a LOT in between. Here are a few of my go-to amari, an excellent place to start if you haven't spent much (or any) time with these liqueurs.

NONINO

● For me, she is the queen. Nonino will convert anyone who says they don't like amaro. It dances away from both extreme ends—bitter and sweet—to something warm, light, citrusy, and herbal. Nonino is lovely and delicate and an ideal after-dinner sip, whether you're drinking it on its own or in a cocktail.

CYNAR

● Cynar is incredibly versatile. A little earthy, a little vegetal, with a balance of sweet and bitter, this bottle plays nicely in many drinks. Try a Cynar Spritz, a Negroni with Cynar subbed in for Campari, or just drink it over ice and mixed with seltzer. Cynar is a great gateway amaro, as it's usually one of the most affordably priced bottles on the shelf.

BRAULIO

● Braulio is a guy in a Shetland wool sweater who's bringing in the firewood before dawn. Piney, bitter, intensely "alpine" (I know it sounds silly, but it'll make sense when you taste it), very clean, and a little bit serious, it brings the outdoors in, in a wonderful way. I sip this neat or over a single large ice cube.

SFUMATO

● Smoky, funky, and unfiltered, Sfumato is a bit like the black sheep of the amaro world—rougher around the edges. If you're interested in natural wine and/or Scotch, you'll love this. Maybe not the place to start on your amaro journey, but absolutely something to try if you've already covered others on this list.

FACCIA BRUTTO FERNET PIANTA

● Slightly more restrained, and far more complex than Fernet-Branca (the most mainstream fernet brand), Faccia Brutto's made-in-the-USA bottle is my favorite version of this style of digestif, and it's the forever star of my Hanky-Pankys (page 261).

BATCHED HANKY-PANKYS

Makes 8 cocktails

12 ounces gin

10 ounces sweet vermouth (I like Cocchi Storico Vermouth di Torino)

4 ounces fernet (I like Fernet-Branca or Faccia Brutto)

Orange twists, for garnish

MAKE-AHEAD: *Up to 6 weeks*

PAIR WITH: *Boozy Chocolate Mousse, to Share (page 239), Dark Chocolate Sheet Cake with Amaro Frosting (page 274), Tahini Hot Fudge Sundaes (page 83), Tre Latti Cake (page 65)*

When I'm drinking amaro on its own, I pour an inch or so into a short rocks glass over a large ice cube. When I'm hosting, I bring out a couple of bottles for people to try. And if I'm feeling extra-fancy, or I want my guests to feel extra-fancy, I make a hanky-panky—a truly underrated classic cocktail that highlights the joys of amari (fernet, specifically). I make a batch of these before a party, stick it in the fridge to chill, and whip it out after dinner. Expert at-home bartending level, unlocked.

In a large pitcher or swing-top bottle, combine the gin, vermouth, fernet, and ⅔ cup water. Stir or gently swirl and chill in the refrigerator overnight. *(The gin mix can be stored in the fridge for up to 6 weeks.)* If you're short on time, pop it in the freezer for an hour or two (a good time would be right before your guests arrive). Just don't forget about it in the freezer—there is water in this cocktail for dilution, so the mixture will freeze.

If you have the space, slip 8 coupes into your freezer, too.

When ready to serve, give the hanky-pankys a last good swirl and pour into the chilled coupes. Express the oils from the orange twists over the drinks, then drop them into the cocktails to garnish. Serve immediately.

A NEW NEW YEAR'S EVE TRADITION

AS FAR AS I AM CONCERNED, there is only one way to do New Year's Eve: at home. Though I have always had a fantasy of attending a New Year's Eve wedding, so if you're planning one, please feel free to pop my invite in the mail. Maybe the reason why a New Year's Eve wedding sounds so perfect is that everyone is gathered to celebrate the same thing. That collective experience is exactly what I crave on the last night of the year—and creating that experience starts with telling people to come over.

At some point, I try to gather everyone for a toast—not one I've prepared, but one we can all make together. I'll ask everyone to come up with a word to describe their year ahead, or to share a sweet memory from the past year, or what they're most proud of. Everyone has a moment to speak, and we all "cheers" each and every answer. Suddenly, we're all celebrating the same things.

As for what we're cheers-ing with: This is a night when I like to stick to just one beverage category: sparkly somethings. I fill the biggest bucket I've got with bottles on ice, and let people serve themselves (and everyone else, too).

Rather than a sit-down dinner, I want my New Year's Eve to be filled with fancy-not-fussy snacks alongside the free-flowing bubbles. I want everyone to mingle around a bubbling pot of fondue. I want to serve caviar, because it's New Year's Eve after all, and if ever there was a time to load up a potato chip with shiny, glittering little black beads, it's now. And when the clock strikes midnight, I want everyone to feel like there's nowhere else they'd rather be.

Menu

For 12+

LOTS AND LOTS OF CHAMPAGNE

CAVIAR SERVICE

WARM ANCHOVY DIP WITH BITTER VEG

THE FUTURE IS FONDUE

DARK CHOCOLATE SHEET CAKE WITH AMARO FROSTING

CAVIAR SERVICE

Serves 6 to 10, easily scaled up

Zest of 1 lemon

1 bunch chives, thinly sliced

½ bunch dill, finely chopped

2 cups (16 ounces) crème fraîche or sour cream

Kosher salt and freshly ground black pepper

Salted potato chips

Ice, preferably crushed

4 ounces sturgeon caviar or trout or salmon roe (see Note)

NOTE: When people say "caviar," they're typically referring to the small, black or dark-green beads that come from sturgeon—not the larger, orange-red beads commonly referred to as "roe." Roe is great, too—you'll see salmon and trout varieties most frequently—and is typically much lower-priced than sturgeon caviar. The taste and look of roe is very different from sturgeon caviar, but the big beads give a satisfying pop, and they look positively festive on a potato chip (or a baked potato, or a latke).

PAIR WITH: *Sweet-Salty Pigs in Blankets (page 189), A Scallop Snack While You Cook (page 233)*

The first thing I will say about caviar is it's so totally okay if you aren't into caviar. Even if it has gotten slightly more affordable in recent years, caviar is still a pricey item, and I understand if the whole idea of splurging on a few ounces of fish eggs isn't for you. But if you're going to do it once a year, do it on New Year's Eve, and drink something sparkling alongside. The perfect bite is a potato chip, a smallish scoop of this dill-y crème fraîche, and a little spoonful of caviar on top. Speaking of: The spoon choice is important here. Caviar absorbs the flavor of metals, so rather than a silver spoon, go for something nonreactive, like mother-of-pearl, if you can.

In a medium bowl, combine the lemon zest, all but 1 tablespoon of the chives, and the dill. Stir in the crème fraîche and season with a tiny pinch of salt (keeping in mind that the chips and caviar are salty) and a few grinds of pepper. Cover and refrigerate for at least 1 hour, and up to 24 hours.

When you're ready to serve, transfer the crème fraîche mixture to a serving bowl, scatter the remaining chives over the top, and set out a bowl of potato chips. Fill a shallow bowl with ice. Use a mother-of-pearl (or other nonreactive material) spoon to transfer the caviar from its tin to an extra-small glass bowl, and set it on top of the bowl of ice. Place the spoon next to the caviar for serving. When the ice begins to melt, drain or dump out the bowl, and refresh with new ice, repeating as necessary until there is no more caviar left.

THE FUTURE IS FONDUE

Serves 8 to 16 alongside more dishes, or 4 to 6 as a main

FOR THE FONDUE

1½ pounds aged Gruyère cheese

½ pound sharp white cheddar, raclette, Comté, and/or Fontina cheese

2 tablespoons cornstarch

1 garlic clove, peeled and halved

2 cups dry white wine (I like a zippy, high-acid bottle like dry Riesling or chenin blanc to enjoy with the fondue later)

1 tablespoon Dijon mustard

2 tablespoons kirsch, brandy, or cognac (optional)

Kosher salt and freshly ground black pepper

FOR SERVING

1 large loaf crusty bread, cubed or sliced

½ to 1 pound sliced crunchy vegetables such as celery, fennel, and/or radishes (optional)

½ to 1 pound steamed, boiled, or roasted fingerling potatoes or cooked tater tots (optional)

½ to 1 pound steamed or blanched broccoli, broccolini, or cauliflower (optional)

3 to 6 ounces sliced or cubed salami, soppressata, mortadella, and/or cooked and sliced sweet or spicy sausage (optional)

2 apples or pears, sliced; 4 fresh apricots, sliced; and/or 1 bunch grapes (optional)

I know you're already envisioning '70s ski lodges and Fair Isle sweaters. But I'm here to tell you: We're bringing fondue back. You don't need any special equipment, you don't need to be fresh off the ski slopes, and you don't need an itchy sweater—in fact, this is a perfect, dare I say sexy, party food.

If you're asking yourself, *Will it fondue?* the answer is yes. Let your creativity run wild with what you want to dip into your cauldron of cheese, but I would recommend at least five options to keep things fun: something bready (I love a cubed crusty sourdough), something potato-y (I love a small yellow potato, steamed, cut in half, and tossed in olive oil and salt), a couple of vegetables (I like something bitter, like radicchio, and something crunchy, like fennel, or broccolini is also fantastic), and a fruit (apple or pear is classic).

MAKE THE FONDUE: Grate the cheeses on the large holes of a box grater or in a food processor fitted with the coarse grating disk. Place the cheese in a large bowl, add the cornstarch, and toss to coat.

Rub the cut side of the garlic clove all over the surface and halfway up the sides of a medium or large heavy-bottomed pot—a Dutch oven works great for this. Add the wine and bring to a gentle simmer over medium heat. Immediately stir in a handful of the cheese mixture. Cook, stirring to prevent the cheese from clumping, until it has melted completely, then stir in another handful of cheese. Continue stirring in the cheese and letting it melt until it has all been added and a smooth mixture has formed. Stir in the mustard and the kirsch (if using) and season with salt and lots of black pepper.

Serve immediately, with any accoutrements you like. Keep the fondue warm, either by serving it directly from the pot over a portable burner or transferring it to a fondue pot, rice cooker, or slow cooker. And if you don't have any of those devices, don't fret! Just bring the pot to the table straight from the stove and dip away. If the fondue begins to solidify on top, give it a good stir and/or slightly increase the heat for a couple of minutes, and/ or pop it back on the stove for a minute or two as needed. *Any leftovers can be stored in an airtight container in the refrigerator for up to 3 days; before serving, rewarm in a pot over medium-low heat or microwave in 1-minute intervals until gooey.*

WARM ANCHOVY DIP WITH BITTER VEG

Makes about 1¼ cups

1 head garlic (about 14 cloves)

Kosher salt

5 tablespoons extra-virgin olive oil

2 teaspoons fennel seed

½ teaspoon red pepper flakes (optional)

1 (6- to 8-ounce) tin or jar high-quality oil-packed anchovy fillets (about 40 fillets), drained

5 tablespoons unsalted butter, cut into pieces

⅓ bunch fresh parsley, finely chopped

Zest and juice of 1 lemon, plus more juice as needed

1 pound chicory leaves, such as radicchio, Castelfranco, Treviso, and/or Belgian endive

1 baguette or other crusty bread loaf, toasted, if desired, and sliced

Lemon wedges, for serving

PAIR WITH: *Fluffy Sheet Pan Focaccia (page 220), Hidden Treasures Salad (page 198), Crispiest Chicken Milanese with Spicy Balsamic Arugula (page 150)*

This is a dip people will smell before they taste: a whole head of garlic, forty anchovies, and bloomed spices. Your guests will start clustering by the dip, positively intrigued. "I don't like anchovies," someone will inevitably say, either in their head or out loud, and then, when convinced to try one bite of this warm dip scooped with radicchio or baguette, that person will immediately change their mind forever. This dip deserves the holiday-party spotlight, but I also like to bring it back during the gray days of January, February, and March, when my dinner table needs a jolt.

Smash and peel the garlic cloves. Sprinkle them with a pinch of salt and use a knife to smash and chop until the garlic is so finely chopped, it's almost a paste. (Alternatively, use a food processor or mortar and pestle—if either tool is having some trouble, add a bit more salt.) Transfer the garlic to a medium saucepan with the olive oil, fennel seed, and red pepper flakes (if using). Cook over medium-low heat, swirling the pan occasionally, until the mixture is very aromatic and bubbling slightly, 10 to 12 minutes—if it starts to brown or bubble furiously, reduce the heat to low.

Sprinkle the anchovies with a pinch of salt and use a knife to smash and chop until the anchovies are finely chopped, nearly to a paste. (You can use a food processor or mortar and pestle for this, too.) Transfer the anchovies to the saucepan and increase the heat to medium. Cook, whisking to emulsify, until fragrant and bubbling, about 3 minutes. Reduce the heat to low and cook, stirring occasionally, until the flavors have melded, 5 minutes more.

Remove the anchovy mixture from the heat. Add the butter, a few pieces at a time, whisking to melt and incorporate. Stir in the parsley, lemon zest, and about 2 tablespoons of the lemon juice. Taste the dip: If it's too salty or too rich, add more lemon juice until it tastes right.

Arrange the chicories on a platter, lightly drizzle with olive oil, and sprinkle with a pinch of salt. Arrange the bread on a separate platter. Serve the dip warm, straight from the pot, or in a heatproof bowl with the lemon wedges for squeezing.

vintage glassware

I PLANNED MY WEDDING IN FIVE WEEKS. Well, I should say, I had to *re*-plan it in five weeks due to my original upstate New York venue not actually being a real wedding venue. (Come over sometime and I'll tell you that story.)

Fatefully, Storm King Art Center—a sprawling sculpture garden, where I never could have dreamed of getting married—was nearby, and happened to have a wedding cancellation on our exact date. I do not recommend planning a wedding in five weeks, but thankfully, it all turned out exactly how it was supposed to. I have so many sparkly memories from that day: our best friend officiating, a disco ball dance floor, a beloved restaurant of ours serving plate after plate of stunning food; looking around at all of our family and friends and watching them cheers each other.

The glasses they were clinking were vintage coupes my mom had spent the prior year sourcing. She drove all over the state of Michigan, where she lives, to find the prettiest glassware. Thanks to her, every guest at our wedding had their own unique glass to sip from. And that was when my obsession with vintage tableware—glassware, in particular—really began.

Sure, buying something shiny and new is its own thrill. But finding a piece of vintage tableware that really speaks to you—and comes with its own history, that you add

to every time you pour a drink, or set the (mismatched) table, or cheers your friends—makes every night at home feel just a little more special. Which is exactly why my mom is still sourcing vintage for me, or really, for Big Night—we stock the shelves with her vintage glassware finds. Here are some of her sourcing tips:

IN PERSON

- Estate sales—typically items are half price the second day and 75 percent off the third day

- Thrift stores & antiques stores—any time you're headed out on a road trip, do a search to see which shops might be located along your route (or worth going out of your way for)

ONLINE

- eBay—If you like an item, check out the rest of the seller's collection, along with their customer reviews.

- Etsy—Tends to be more expensive than eBay, but you get to skip the auction.

- If you find a vintage item you like, learn a little bit about it so that you can search for more like it by using keywords, like "Vaseline glass," "silver fade," or the name of the maker.

DARK CHOCOLATE SHEET CAKE WITH AMARO FROSTING

Serves 16 to 24

Growing up in Texas gave me many gifts. One example: Texas Sheet Cake, which I learned about from my friend Kate and her family from Houston (another one of those gifts). This one is inspired by their famous version, with the addition of an amaro icing, which feels especially luxe for a holiday party. As its name suggests, the cake is made in a sheet pan, so you can easily slice it for as many people as you have. But fair warning: If you leave me (or anyone else) alone with this thing . . . I cannot guarantee everyone will actually get a slice.

FOR THE CAKE

Nonstick cooking spray

3 cups all-purpose flour

2⅔ cups granulated sugar

1 teaspoon baking powder

1 teaspoon baking soda

2 teaspoons kosher salt

¾ cup (1½ sticks) unsalted butter, cut into pieces

3½ ounces bittersweet chocolate, chopped

6 tablespoons unsweetened cocoa powder (preferably Dutch-process), sifted or whisked to remove any lumps

¾ cup buttermilk or plain full-fat yogurt

3 large eggs

3 tablespoons neutral oil

2 teaspoons pure vanilla extract

Flaky sea salt, for serving

FOR THE ICING

10 tablespoons (1¼ sticks) unsalted butter, cut into pieces

½ cup plus 2 tablespoons unsweetened cocoa powder (preferably Dutch-process), sifted or whisked to remove any lumps

½ teaspoon kosher salt

5 to 7 tablespoons medium amaro, such as Averna or Montenegro

1½ teaspoons pure vanilla extract

2½ cups powdered sugar

MAKE THE CAKE: Position a rack in the center of the oven and preheat to 350°F. Coat a 13 × 18-inch rimmed baking sheet well with cooking spray, ensuring that all the sides and corners are coated, then line with parchment paper.

In a large bowl, whisk together the flour, sugar, baking powder, baking soda, and salt.

In a medium saucepan, combine the butter and chocolate. Add 1½ cups water. Melt over medium heat, stirring occasionally, until the mixture is smooth. Whisk in the cocoa powder until smooth.

Pour the chocolate mixture into the bowl with the dry ingredients and whisk to combine, reserving the saucepan to use for the icing (no need to wash it).

In a medium bowl, whisk together the buttermilk, eggs, oil, and vanilla. Slowly whisk the egg mixture into the chocolate mixture until the batter is smooth. Pour the batter into the prepared pan (it will be fairly loose).

Bake, rotating the pan halfway through, until the cake is starting to pull away from the edges of the pan and a tester inserted into the center comes out with only a few crumbs attached, 27 to 30 minutes. Transfer the cake (still in the sheet pan) to a wire rack to cool to room temperature, at least 30 minutes. *Store the unfrosted cake at room temperature, tightly wrapped, for up to 2 days.*

MAKE THE ICING: In the same saucepan you used for the chocolate, melt the butter over medium heat. Reduce the heat to low and whisk in the cocoa powder, salt, 4 tablespoons of the amaro, and the vanilla. Remove from the heat, then whisk in the powdered sugar, ½ cup at a time, until smooth. Taste the icing and, if you'd like a more pronounced amaro flavor, whisk in the remaining 1 to 3 tablespoons amaro.

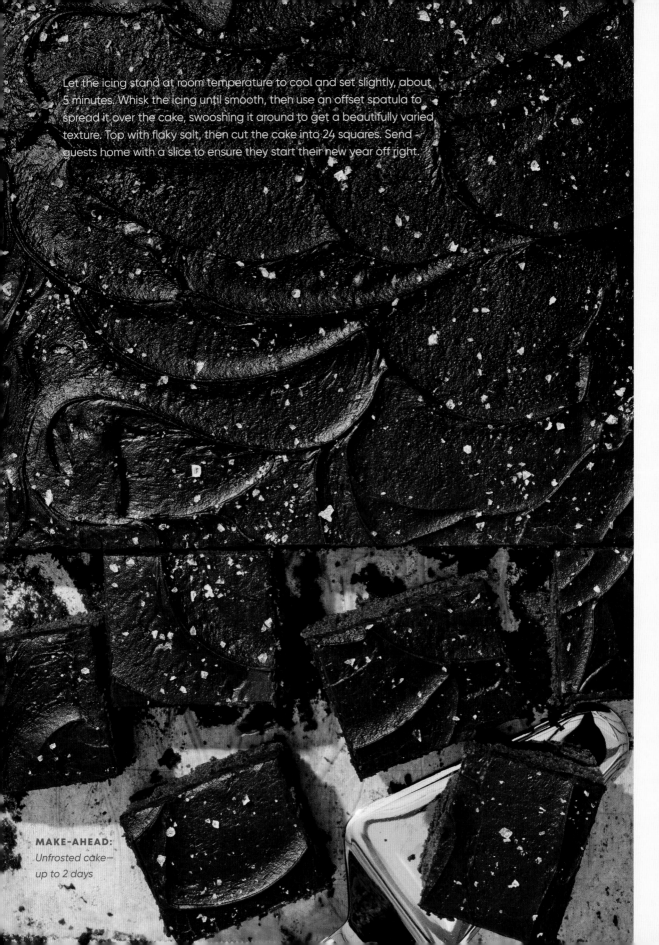

Let the icing stand at room temperature to cool and set slightly, about 5 minutes. Whisk the icing until smooth, then use an offset spatula to spread it over the cake, swooshing it around to get a beautifully varied texture. Top with flaky salt, then cut the cake into 24 squares. Send guests home with a slice to ensure they start their new year off right.

MAKE-AHEAD:
Unfrosted cake—
up to 2 days

little tips for big nights

Big Night Math

2 ounces of cheese per person for a cheese plate	**3 to 5 cheeses** for a cheese plate	**7 spritzes** per bottle of sparkling wine	**2 ounces** amaro, **4 ounces** sparkling, **1 splash** seltzer for any spritz
1 pound of pasta per 4 people	**8½ minutes** of boiling for jammy eggs · · · · · · · · · · · · · **11 minutes** for hard-boiled/ deviled eggs	**2 hot dogs,** minimum, per person	**1 hour before,** pull cheese out of fridge
1 bottle of wine **yields 5 glasses** 	**1 bottle of wine** per person *(just in case!)*	1½ pounds Thanksgiving **turkey** per person	4 to 6 ounces **bacon** per person
½ pound **clams** per person	1 extra pint of **ice cream,** for backup	**3 olives** per martini	**10 minutes** for a Social Nap

More Menus for More Big Moments

WHEN YOU'RE MEETING YOUR PARTNER'S PARENTS

Ricotta Toasts for
Every Mood — 44

Party Chicken
with Feta & Fennel — 27

Apple & Miso Cobbler — 175

DINNER WITH FAMILY YOU HAVEN'T SEEN IN FOREVER

Amatriciana for a Crowd — 219 *or*
Crispiest Chicken Milanese with Spicy
Balsamic Arugula — 150

Family-Style Crème Brûlée — 256

A BIG NIGHT TO TURN AROUND A BAD DAY

Creamy Tomato Soup — 224 *and*

Focaccia Grilled Cheese — 225 *or*
DIY BLT Night — 86 *or*
Big Calzone Night — 157

Sundae Bar — 83

YOU JUST GOT A PROMOTION

Caviar Service — 266

A Scallop Snack
While You Cook — 233

The Future Is Fondue — 269

Batched Hanky-Pankys — 261

LONG LUNCH OUTSIDE WITH FRIENDS YOU HAVEN'T SEEN IN FOREVER

Hidden Treasures
Salad — 198

Clam & Corn Pasta — 141

Corn & Strawberry
Pop-By Cake — 104

DATE NIGHT TO IMPRESS SOMEONE NEW

Marinated Olives
for Anytime — 80

A Scallop Snack
While You Cook — 233

Slow-Roasted Shawarma-
Spiced Salmon — 75

Boozy Chocolate Mousse,
to Share — 239

A Seasonal Sides Matrix

So you've figured out what you're serving for dinner. But what's happening on the side?

You might know what your star of the dinner show is, but you need supporting actors to round it out. And for whatever reason, those can sometimes be the hardest roles to cast.

Consider this matrix your Magic 8 Ball, here to give you answers when you are completely lacking in side-spiration.`

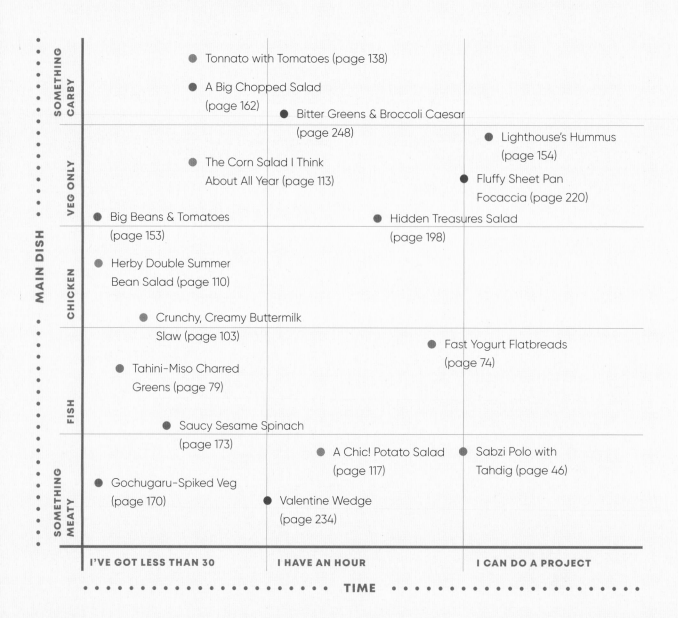

MAIN DISH

SOMETHING CARBY
- Tonnato with Tomatoes (page 138)
- A Big Chopped Salad (page 162)
- Bitter Greens & Broccoli Caesar (page 248)

VEG ONLY
- Lighthouse's Hummus (page 154)
- The Corn Salad I Think About All Year (page 113)
- Fluffy Sheet Pan Focaccia (page 220)
- Big Beans & Tomatoes (page 153)
- Hidden Treasures Salad (page 198)

CHICKEN
- Herby Double Summer Bean Salad (page 110)
- Crunchy, Creamy Buttermilk Slaw (page 103)
- Fast Yogurt Flatbreads (page 74)

FISH
- Tahini-Miso Charred Greens (page 79)
- Saucy Sesame Spinach (page 173)

SOMETHING MEATY
- A Chic! Potato Salad (page 117)
- Sabzi Polo with Tahdig (page 46)
- Gochugaru-Spiked Veg (page 170)
- Valentine Wedge (page 234)

| I'VE GOT LESS THAN 30 | I HAVE AN HOUR | I CAN DO A PROJECT |

TIME

Back-Pocket Sauces, Dips & Dressings + How To Use Them

TOMATILLO-AVOCADO SALSA

- Braised al Pastor-ish Tacos (page 62)
- The Corn Salad I Think About All Year (page 113)
- In breakfast tacos
- To pour over a pan-seared or grilled piece of fish

RANCH-ON-EVERYTHING DIP

- Crunchy, Creamy Buttermilk Slaw (page 103)
- Substitute dressing for Herby Double Summer Bean Salad (page 110)
- To go alongside Sweet-Salty Pigs in Blankets (page 189)
- A dip for takeout pizza

TAHINI-MISO SPECIAL SAUCE

- Tahini-Miso Charred Greens (page 79)
- On lamb burgers
- Thinned to make salad dressing

SPICY BALSAMIC

- Crispiest Chicken Milanese with Spicy Balsamic Arugula (page 150)
- Toss into warm roasted vegetable

TONNATO

- With Tomatoes (page 138)
- As a dip on its own, for crudités or Fritto Misto (page 97)
- Substitute for mayo on BLT Night (page 86)

LIGHT & BRIGHT CAESAR

- Bitter Greens & Broccoli Caesar (page 248)
- As a side sauce for Crispiest Chicken Milanese (page 150)
- Substitute dressing on Crunchy, Creamy Buttermilk Slaw (page 103)
- Substitute for dressing in A Chic! Potato Salad (page 117)

Dress Up Vanilla Ice Cream

- Good olive oil + flaky salt
- Crumbled halva
- Crumbled Oreos
- Chili crisp of your choice
- Tahini Hot Fudge (see page 83)
- Quartered strawberries, tossed in a bit of sugar

ACKNOWLEDGMENTS

Big Night (the book, and the shop) would not be Big Night without these people:

To my mom, Anne. Thank you for everything, especially your endless passion and incredible eye for vintage, which made this book so beautiful. Every time I see the glassware, the platters, the deviled egg plate, I think of you.

Rebecca Firkser, I simply could not have created this cookbook without you. Thank you for diving headfirst into this project, on a frankly bananas timeline, and bringing your deep well of expertise, knowledge, organizational skills, and creativity with you to make every single recipe better.

Mia Johnson, I want to yell your name from the rooftops! Thank you for creating the visual identity for our shops, our brand, and now, this book. I am forever grateful for your understanding of the perfect shade of red, for your iconic illustrations, for that tile checkerboard, and so much more. I couldn't feel luckier to work with you on all of it.

Arden Shore, thank you for that pivotal brainstorm on your couch. I hadn't yet put a single word down on the page, because I was scared, and you knew it. The next morning, I woke up and started writing. That's because of you.

Shooting this book was, without a doubt, one of the most thrilling creative experiences of my life, thanks to these women:

To my photographer, Emma Fishman, thank you for saying yes to doing this book the very first time I met you. Thank you for the late-night fun in the studio, when we just wanted to get that martini shot right. Thank you for making our eight-day, no-breaks marathon an experience I'll treasure forever.

To my food stylist, Monica Pierini. It was an honor to watch you make these recipes more beautiful than I could have ever imagined. Your kindness, calm, strength, organization, and vision on set made all these photos possible.

To my prop stylist, Stephanie DeLuca. Thank you for extending the Big Night world with me onto the pages of this book, and for (gently) reining me in when I kept wanting to add, add, add more to the shot. Your visual editing is unmatched. And thank you to Maggie Nemetz for assisting on props. Having you there meant so much.

To Stacey Mei Yan Fong, Ariella Sperling, Claire Trice, and the entire Big Night team: Thank you for holding down the fort, quite literally, during the making of this book. I was only able to create this book because you all stepped in, stepped up, and led our shops when I was deep in the tunnels of writing, editing, and shooting. Your support and encouragement kept me going.

To our customers. You are the most important part of Big Night, and I wrote this for you. Thank you for making my dream of creating this shop—and this book—possible.

To my editor, Amanda Englander. Thank you for your cold email, for secretly taking your boss shopping at Big Night, and for seeing—like, really seeing—what this book could be. Your willingness to bet on me gave me the confidence to write this. Thank you, also, to the whole team at Union Square for your dedication to this book: Caroline Hughes, Lisa Forde, Renée Bollier, Ivy McFadden, Kevin Iwano, Blanca Oliviery, and copyeditor Terry Deal.

To my book agent, Sarah Smith. I love working with you, but more than that, I love knowing you.

Thank you for helping me keep a level head and for being my support and advocate during this entire process.

Thank you to Roya Shariat, Stacey Mei Yan Fong, Naama and Assaf Tamir, and Christine Collado for sharing your recipes, words, talents, and knowledge. Your contributions are the essential icing on the cake of this book.

To the friends who, for years, have come over, sat around our hot pink table, and never once complained that dinner wasn't ready until 10 p.m. This book is inspired by our time around that table together: Kate Nemetz, Noelle Lyles, Skyler Okey, Anya McCall, Alex Van Trigt, Julianna Stamos, Hannah Waitt, Charisse Thompson, Erica Padgett, Chenaya Devine Milbourne, Leanna Balaban, and Sarah Ferrie.

And Kate, thank you for (at the time of this writing) ten years of Ham Party. It's what convinces me, year after year, that nothing is more fun than hosting at home.

Thank you to Jordan Haro and Ann Lupo, for your boundless enthusiasm for all things Big Night, and for documenting this ride with me at every stage.

Thank you to my unwavering crew of recipe testers: Anya McCall, Alex Van Trigt, Julianna Stamos, Arden Shore, Jess Basser Sanders, Julia Yasser, Caleb Pershan, Maggie Nemetz, Kate Nemetz, Nancy Nemetz, Zoe Nemetz, Victoria Soto, Skyer Okey, Carlo Mantuano, and Nick Rizzo.

To my childhood friends who are family, whose houses have always felt like homes: Thank you to the Stites, the Waitts, the McGhees, and the Nemetzes for always welcoming me into yours.

To Kellie and Len, for all your love and support, and for letting me cook you Thanksgiving in July.

To my grandmother, Mary Virginia Kane, the consummate host. You always made every single person feel welcomed, loved, and included. Thank you for showing me the way.

To my dad, Robin. I may have inherited your tendency to "pack it all in" (see page 32), but I am grateful for it. Thank you for showing me what ambition and hard work look like, and for always being in my corner.

To my sister, Serena. Thank you for cooking Persian food with me, and for being the life of any and every party. You make every Big Night even better.

To Skyler Okey. You are Big Night's secret sauce. You are the best cheerleader, listener, decide-by-doing-er, and life partner I could ever ask for. Thank you for believing in me, and in Big Night, from the beginning.

To my husband, Alex, to whom this book is dedicated. So much of our life together is in these pages, and our future is even brighter. Thank you for making me my perfect martinis, and for every other cocktail in this book. Thank you for encouraging me to be more like me—always. Your support in every single moment of this journey is what got me here. I love you.

INDEX